A New Theory

of

THE UNIVERSE

By Paul Tatham

Published by Lulu.com

Published August 2020

© Paul Tatham 2020.

ISBN 978-1-291-37200-7

All rights reserved. No part of this publication may be reproduced, stored in a retrieval system, or transmitted in any form or by any means, electronic, photocopying, recording or otherwise, without prior written permission of the publisher.

CONTENTS

Chapter	Page
1. A summary of my theories.	9
2. Einstein's Relativity is shown to be nonsense.	13
3. The misunderstanding of mass and momentum.	43
The Higgs Theory	
4. The misunderstanding of gravity and weight.	53
A quantum solution for gravity.	
The 'proof' of General Relativity is unreliable	
Gravitational waves	
5. The misunderstanding of Time.	67
Why clocks actually do go slow at high velocity.	
6. The misunderstanding of waves, particles and reality.	75
7. The Fabric and Size of Space.	83
8. Dark Matter, Dark energy and Galaxies	91
9. What is Energy?	101
The structure of waves and charged particles	
10. My Theory of 'Conserved Relativity'.	115
The Contraction of the Universe.	
Why time will stop and space disappear	
The Big Bang, and Black Holes	
11. My Theory of the Creation of the Universe.	129
12. Commercial Opportunities of Gravity.	141
How to make objects weightless.	
Is it possible to extract energy from gravity?	
13. How much of Science Fiction is possible?	145
14. Possible Explanations for ESP.	149
Telepathy and Astrology	
15. My process for finding the solutions	153
16. The reasons I did all this.	159

About the author.

Paul Tatham is an engineering graduate with a practical and common-sense outlook, definitely not an academic, and is now retired after a number of years in Xerox Corporation and the computer industry. He brings logic, common sense and a practical approach to the understanding of the universe.

This book is totally the work of the author, Paul Tatham, and the contents, ideas and conclusions contained in it do not represent, and are not connected in any way to Imperial College, London University.

Introduction

This work began as a retirement project simply because I was interested in gravity but I had never read Einstein's theories. The project became larger and larger and eventually turned into a book. I read about the current understanding of gravity and the universe but I was not entirely convinced due to my background in practical engineering and my common sense attitude and logic. Two things immediately struck me as odd.
1. Why has curved space become totally accepted as being the source of gravity when it is not a quantum solution it is just a theory? How can it be said that space is curved if the fabric of space itself has not been defined? Light lensing is said to prove that space is curved, but graviton radiation from protons will attract light waves and produce exactly the same light lensing, so does Einstein's theory really have any proof?
2. Mass has never been properly explained. Its properties are clearly defined and we all use it in school physics, but what is it? If the mass of something increases it should be possible to observe a change, such as; it gets bigger, or it changes colour. Mass must be 'stuff' for it to have a property. How can it be said that the Higgs Boson gives mass to everything if what mass is, has not been defined? Surely resistance to motion is not caused by external factors, it is a property of a charged particle itself.

Curved space is Einstein's theory of General Relativity (GR), and GR was derived from Einstein's Special Relativity (SR), so I spent some time reviewing Einstein's SR and found two errors.

Those errors are described in chapter two, and because they make SR nonsense, it means Einstein's curved space gravity is nonsense.

So, I came up with a new theory of gravity based on the Standard Model's Graviton, and I describe that theory in chapter four. The theory achieves everything that is used to prove curved space is correct, such as light lensing.

Einstein's error in Special Relativity was that he thought an absolute reference frame of 'stationary' could not be defined, and this led to his creation of inertial frames, postulates and invariant Lorentz Transformation that are unnecessary because stationary is simple to define.

So, Einstein's final conclusion that Lorentz equations represent laws of physics is incorrect. Lorentz is not invariant, they are simply expressions on how stationary observers see a moving object.

I also explain in chapter two why all the sketches that are drawn to explain Special Relativity are wrong because they do not show how the moving observer sees the light pulse. They are drawn to reflect Einstein's mathematics, that a moving observer would see things to be the same as for the stationary observer. I explain why relativity is a completely unnecessary theory and has caused a massive delay in our understanding of the universe.

I also studied mass to determine what is the material that produces the properties of mass that we all know so well, and the answer is that it is the electromagnetic wave that is produced by every charged particle, and that includes a light wave. Mass is explained in chapter three.

Having solved what mass is, I concluded that Einstein's $E = mc^2$ is incorrect! It should be $E = mc^2 f$. Where f is frequency. I had trouble finding anything that Einstein got right. Genius??

When such errors form the basis of later theories, the errors become multiplied a hundred-fold, and one can no longer be certain what is right and what is wrong. Black Hole theory is one example. If my understanding of gravity is correct, then all Black Hole theory is wrong.

Being an engineer, I need to understand the physics of how things work. I cannot simply accept a theory based solely on 'mathematical proof', I must be satisfied that the practical physics is correctly identified and proven, and that is the problem with gravity. The requirement for sound physics, combined with common sense and logic, is where I began this project.

From my basis of gravity and mass it was a simple matter to understand Dark Matter and Dark Energy and these are described in chapter eight.

You will find many new theories elsewhere in the book such as how to reduce gravity and harness its energy, why the universe is expanding, and why it must eventually contract back into a dot and cause another Big Bang. I also explain Time because it is misunderstood.

The basic errors of Einstein's Relativity, Mass and Time explains why such little progress has been made in understanding the universe, but I don't believe anyone can say any individual theory is correct unless it can help to answer everything in the universe so that it all fits together, and that is why I have produced a summary of everything in chapter one.

Some brilliant work was done many years ago by Newton, Rutherford, Planck, Maxwell, and Hubble, plus Einstein's $E = mc^2$ (which should actually be $E = mc^2f$) and their achievements are all that is needed to understand the universe, and once understood, it is no more complex than school level physics.

I show that a field in space is required for the universe to function, and that this is simply radiation from a proton as it decays. (It is not a Higgs Field). These waves are the basis of most of the weird things that happen, including gravity.

Most of my theories in this book are new and significantly different from those of current scientific opinion, and perhaps it is the errors that have led to many scientific theories that I find to be absurd, such as multiple universes and multiple dimensions. These are purely theoretical physics based on mathematics and are not physically possible.

Little definitive progress in solving gravity has been achieved in 100 years whereas we should be extracting clean cheap energy from gravity by now, and I show in chapter 12 how this might be done. I am continuing to try to get a paper published to prove Einstein's theories are unhelpful, otherwise gravity will only be understood if it occurs by accident. Spending another 100 years looking at ground state fluctuations, vacuum energy and branes is not going to give a solution because there is no force that can bend the electromagnetic fabric of space.

I have done my best to clarify how I believe the universe works. I hope you enjoy the book, and thank you for reading it.

Paul Tatham.

Chapter One

A SUMMARY

In my analysis, some fundamental errors and misunderstandings have occurred in many areas of physics, and these are explained in this book.

While most of Special Relativity (SR) is simple and obvious, Einstein did not realise that the absolute reference frame of 'stationary' could be defined, and unfortunately that led to a string of incorrect ideas, statements and unnecessary mathematics, which in turn led to equations and conclusions on time and length that do not represent laws of physics.

When stationary is defined and forms an absolute reference frame there is no requirement for his postulates, and there is no conflict between physics and relativity, but he also used relativity assumptions, and not laws of physics, in his mathematics so that his conclusions remain as statements of relativity and not physics. There actually is no requirement or justification at all for his theory of Special Relativity, and it has wasted about 100 years in lack of progress. It does not really matter that Special Relativity is incorrect, but it is critical to GR and the validity of the theory of gravity.

General Relativity is a geometrical concept that is impossible in terms of real physics, and the two reasons Einstein had for developing the theory (the inaccuracy of Newton's equation over large distances and speeds, and acceleration-gravity equivalence) are all easily solved by realising that the gravitational force is an electromagnetic wave as discussed below. The wave that creates Newton's gravity, *causes* Earth to rotate, it is not that Earth's rotation, by an unknown fictitious force, plus his theory of relativity, somehow causes space to curve.

Einstein's theory that curved space creates gravity is impossible from a physical viewpoint. Space has dimension therefore it is not 'nothing'. It is filled with material, and that can only be electromagnetic waves. These waves travel at a fixed velocity and cannot contract, but even if the waves could be curved, an object with momentum would not follow such curved electromagnetic waves, it would continue straight. But the GR theory is not the source of gravity because Einstein's SR is incorrect.

If one accepts that the laws of physics must apply in all situations, then a photon of light must have mass because it is energy, and that means every electromagnetic wave has mass, including that created by a charged particle. Maxwell showed that an electromagnetic wave is self-propagating. It is momentum, so it is the mechanism of mass. And if an electromagnetic wave has mass and momentum, then a charged particle has no mass. It is its wave that is mass and momentum, and the wave 'carries' the particle. The mass is the size of the wave, and that is Planck Length x wave width.

Resistance to motion is not caused by the Higgs Field and is not directly due to mass, it is the delay in the process of a charged particle absorbing energy and converting it into a wave of momentum that is mass, which then allows it to have velocity by means of the self-propagating wave. This process must be understood if gravity is to be understood. The Higgs Field did not give mass to everything.

In my analysis, gravity is caused by radiation from protons in atoms. The Standard Model suggests that a particle called a graviton is the cause of gravity. We know that protons decay releasing w and z bosons and radiation, so I have concluded that my solution of proton radiation is the hypothetical graviton. Gravity is due to a transfer of such radiation from protons in one atom (momentum of an electromagnetic wave), to electrons in a second atom, and because the polarity of the proton wave, the graviton, is opposite to that required by an electron, it causes the electron, which has no resistance to motion, to travel in the opposite direction to the proton wave. Thus, radiation from the protons in one atom pulls the electrons in a second atom, and those pull the nucleus and atom.

All radiation from protons in a spherical object will balance to produce a vertical force with momentum from all areas of the surface. All massive objects are spherical because of this gravitational force.

Once the second atom is moving towards the proton / boson radiation, the electrons receive blue-shifted waves, thus absorption is in larger quantities of energy, and the electron accelerates.

That is the simple basis of why gravity produces acceleration that Einstein was trying to resolve with General Relativity. And if, in Newton's equation for the force of gravity, distance r is replaced by wave velocity, times time, the problem of inaccuracy over long distances is solved.

A charged particle simply absorbs the energy of a wave and converts it into its own wave. That is momentum transfer, and the source of the gravitational force. The particle does not behave as a wave. It has no duality. It is simply carried by its field wave. The field produces duality.

So, there is a misunderstanding of waves, particles, quantum mechanics and reality. If an electron is passed through two slits, the wave passes through both slits whilst the electron passes through just one slit. The two parts of the wave interfere after the slits, but the interfered wave is the only momentum that the electron can have, so it must remain within the part of the wave that still exists, thus it must produce an interference pattern on a screen as if it behaved as a wave.

Thus, if a charged particle does not wave or have mass, quantum mechanics and wave-function mathematics is of doubtful validity and the concept of probabilities and 'realities' seems to be incorrect. It is the many strands making up the waves that allow quantum computers to work so fast.

The Uncertainty Principle is a measure of position and momentum, but this is two parts of a directly related electron system. If energy is used in the measurement it will affect the wave, and that will affect the particle, thus it is impossible to measure position and momentum at the same time, and entanglement, where two photons appear to communicate faster than light, can be explained by the existence of a third energy of proton radiation.

The radiation from protons is more than just the source of gravity. It emerges from every massive object in trillions, and expands when released from the confinement of the massive body. It is the material that gives space its dimension. It is being created all the time, carrying the galaxies, and the whole universe outwards. It is Dark Energy.

Because it is being created and is expanding, it carries galaxies further and further apart. But there is a boundary beyond which radiation from protons, and light waves, have not yet passed, even after billions of years at the speed of light. Beyond that boundary there are no waves. There is no material to form the dimension of space. Therefore, there is no dimension and the location does not exist.

The proton radiation is mass because it is an electromagnetic wave. It is probably Dark Matter. It has no charge because it is a wave, not a particle. It is really 'Dark Wave'. Further away from its source where its density and gravity is weaker, it becomes Dark Energy (sometimes referred to as 'anti-gravity'), expanding the universe. It is the missing mass that makes up 75% of the universe. But it is not what holds galaxies together. That is gyroscopic precession.

All stars and many planets will spin because proton waves (gravity) only pull electrons and these can only be pulled in a way that causes the star to spin. This spin, and gravitational forces causes precession, and precession holds the galaxies together.

Gravity created by proton radiation satisfies the strange orbit of Mercury because it causes Mercury to rotate. This creates precession along its orbit around the sun. The rotation and precession causes Mercury to stay in the sun's orbit for longer than gravity alone would permit. All of the properties of gravity that are attributed to General Relativity are equally explainable by radiation from protons, such as light lensing and redshift.

Time is misunderstood. It is not a dimension. It has no 'speed'. It cannot dilate. It is just a mathematical calculation for the use of energy in an event. But all these events are completely independent, so there is no universal time, nor any concept of time that links them all together into a single dimension or continuum. Time is only a mathematical tool that we have invented to allow us to measure the rate of change, or velocity of an event. It is simply time = distance / velocity and nothing more. It has no role in the universe. Time does not *cause* anything to happen. There is no 'past', there is only 'the way things were before the event occurred'.

A clock is just a machine that will go slower at high velocity. Energy is added to achieve the high velocity and this increases the mass of all the components in the clock, and that means more energy is required to maintain motion. If that extra energy is not supplied, the clock will slow.

When all protons everywhere have released all of their energy, space will disappear and the universe will lose its dimension. Every item of energy or matter will quickly collapse back into a dimensionless dot, and a new Big Bang will occur. Such a cycle may be repeated forever. The Big Bang that we know about, was probably not the first.

Chapter Two

HOW TO MAKE PROGRESS IN UNDERSTANDING THE UNIVERSE? SCRAP EINSTEIN'S RELATIVITY!

Scrap everything done by Einstein. Forget length contraction, time dilation, space-time, curved space and anything else that seems a bit strange. All of it is nonsense and I will explain why. There is a danger that these incorrect laws of physics are becoming accepted when there is not a single shred of physics proof. Light lensing is not conclusive proof that space is curved. Einstein's theories have already wasted over 100 years in achieving progress in the understanding of the universe and it is time to put that right.

All of the facts that have been observed, or proven in experiments that have been attributed to Einstein's theories, such as gravity, light lensing, Mercury's orbit and the behaviour of Muons, can be explained more simply in other ways and I have done so in later chapters.

I spent considerable time attempting to understand why his theories of relativity are not valid, but it becomes a 'can of worms'. The deeper you dig, the more you find cannot be right and so you have to go right back to where Einstein began because that is where he went wrong.

I am a practical and logical thinking engineer and so I focus on facts, not theories, but Einstein's relativity is all mathematics. There is not a single fact of physics to support it. I immediately ruled out some of the amazing conclusions reached by Einstein, such as time dilation and length contraction because they are impossible and there is no proof.

I will explain why clocks go slow at high velocities, and why muons do odd things. There is a practical and sound explanation for everything and I will explain everything later in the book. But I almost gave up attempting to explain why Einstein is wrong.

I originally started this chapter a long time ago by explaining each step why Einstein is wrong, but gave up because it is all confused nonsense. It is not possible to go through any individual statement or equation and say that this is right and that is wrong. In the end I realised that it is the entire idea of Einstein that is wrong. His train of thought is wrong from the beginning. And you cannot find his error just by reading his paper, you have to read his entire book to understand his (incorrect) train of thought.

It is impossible for scientists to find a quantum solution for Einstein's gravity because there isn't one. It is also impossible to find out what Dark Energy and Dark Matter are if you believe in Einstein's theories. You will also have to re-consider $E = mc^2$ because that is only half correct, and I will explain all that later in the book.

Most of relativity is easy to understand and is simply caused because of the time light or sound takes to travel from one observer to another some distance away. It is obvious that two such observers will disagree on the exact time that an explosive 'bang' occurred.

The work by Lorentz is brilliant. It is an excellent piece of mathematics that enable you to calculate how someone in a moving system will perceive an event that occurs in a stationary system, and vice-versa. But it is all visual perception. It is how things 'appear', not the reality of how things are, and any conclusions reached by using the mathematics will be unreal, because they are based on perceptions caused by the speed of light.

But the correct way to understand things is to concentrate on reality rather than visual perceptions. I explain how to define 'stationary' and from that solid base, the speed of light and all other velocities can be accurately defined.

The only reason for discussing relativity at all is to suggest that General Relativity and curved space are not the cause of gravity. Then in chapter four I will suggest what I think is the cause of gravity. Special Relativity is unimportant except that it forms the basis of General Relativity and gravity.

I had decided at the start of this project that curved space is impossible. Space must be filled with electromagnetic waves because it has dimension, and these waves travel at the fixed maximum velocity, and nothing can contract or bend those. And even if they could be bent, they do not have the solid strength to cause a passing object with mass and momentum to curve, and so curved space cannot be the source of gravity. I also decided fairly quickly, what I thought the cause of gravity was, and I offer my solution in chapter four.

So, my purpose in looking at General Relativity (GR) was simply to understand how the concept of 'curved space' came about. But after looking at it briefly, the whole basis of GR comes from SR and in the understanding that length actually contracts as velocity increases, so that is where I began this project.

Einstein's Paper known as Special Relativity. (SR).

This should be mandatory reading in schools. It is a classic, as it is such a concoction of clocks, moving rods and observers.

Einstein followed the wrong train of thought from the outset in his book 'Relativity, the Special and General Theories', and in his paper, 'On the Electrodynamics of Moving Bodies' and that led to too many things going wrong.

There was sufficient doubt about Einstein's theory and the statements made in it that a complete review of his theory was felt to be appropriate.

The mathematical paper on relativity produced by Einstein and the logic he applied in his book, were reviewed in detail and were found to contain important errors such that he did not prove his postulates to be correct.

It is shown that an absolute reference frame can easily be defined so that the train of thought adopted by Einstein including inertial frames, postulates and ultimately invariant Lorentz Transformation, was not necessary. It was because of his postulates that he believed Lorentz Transformation is invariant, but the postulates are wrong, so his invariant transformation is wrong, which means that his conclusions were just expressions of relativity, not laws of physics.

The graviton proposed in the Standard Model is the basis of a more realistic theory of gravity. This alternative gravity satisfies all of the observations and experiments that are used to demonstrate that Einstein's theory is correct. Such a theory of gravitation is discussed in chapter four.

Einstein's introduction to his theory.
Einstein's theory began with a discussion of electrodynamics, in which he states,

No properties of observed facts correspond to a concept of absolute rest.

And that for all coordinate systems for which the mechanical equations hold, the equivalent electrodynamical and optical equations hold also,

And the further assumption,

That light is propagated in vacant space, with a velocity c which is independent of the nature of motion of the emitting body.

It is will be shown that Einstein's first two statements above are incorrect, but Einstein developed his thoughts and believed that an electromagnetic wave (light) required us to change our understanding of time, and that also implied distance.

His thought process was that because of the delay in communication caused by the speed of light c, we should not assume that time is the same everywhere. He gave examples of this opinion in his Relativity of Simultaneity. He stated,

"Events which are simultaneous with reference to the embankment are not simultaneous with respect to the train, and vice versa. Every reference-body (co-ordinate system) has its own particular time; unless we are told the reference-body to which the statement of time refers, there is no meaning in a statement of the time of an event".

This led him to think that if time is a variable depending on the distance from an event, then it must be different between two observers, because the images from them are events, particularly if one is moving relatively to the other. This led him to think that if one observer is moving relative to the other, then the speed of light c, must be different, because he assumed that there is no absolute reference frame.

The logical steps Einstein used to construct his Special Relativity Theory.
The following are the steps Einstein took to further demonstrate his opinion.
1. Absolute Reference Frame. Einstein accepted Galileo's conclusion that an absolute reference frame could not be defined and that everything moves relatively to each other.
2. Inertial frames. Einstein constructed the concept of inertial frames in order to compare the motion between two bodies. Einstein decided that because an absolute reference frame could not be defined, so that there is no definitive point from which to measure the speed of light, the laws of physics must apply equally in every inertial frame, so that the speed of light must be measured from the frame of every inertial frame itself. That means that the speed of a light-wave must change as it passes through different inertial frames, and this idea led Einstein to define his postulates.
3. The addition of velocities. Einstein's 'Theorem of Addition of Velocities' gives the example; If an observer walks at velocity w, towards the front of a train moving at velocity v, then the velocity of the walking observer, as seen by an observer on the embankment is w + v. He then applied the same approach to light, and because he believed that laws of physics must apply in each inertial frame so that the speed of light in the train is c, then when his addition of velocity theorem is applied it means the speed of light in the train, as seen by an observer on the embankment, is less than c. Or more precisely it is c – v. From this, he drew up his postulates.
4. The Postulates. Because he felt that stationary could not be defined, he created two postulates that he believed must apply to all inertial frames, and these are,
    A. The laws of physics are the same in all inertial reference frames.
    B. In any inertial reference frame, light is always propagated with a definite velocity c that is independent of the state of motion of the emitting body.
Einstein called these his 'principles of relativity'.

5. <u>Conflict between relativity and physics.</u> Einstein's addition of velocities theorem and his postulates produced a conflict between the laws of physics (the light-waves) and his 'principles of relativity, that Einstein had to resolve.
6. <u>Mathematical approach.</u> Because he believed an absolute reference frame could not be defined he believed he was unable to remove the conflict via the laws of physics, and so Einstein decided that relativity had to be the approach to use, and to use the mathematics of relativity to prove his postulates to be correct, and so resolve the conflict.
7. <u>Assumptions in mathematics.</u> It was necessary for Einstein to apply assumptions, and for these he used the same approach as used in his theorem of the addition of velocities as described above.

Review of Einstein's steps.
In step 1. Einstein did not challenge Galileo's opinion that it was not possible to define an absolute reference frame, and that was a fundamental error of omission in his theory. But the law of constancy of light waves, shown by the Mickelson-Morley experiment, and from Maxwell's work, light is self-propagating with the constant speed c, regardless of whether the source is moving or stationary, means that stationary can be easily defined, as follows.

The speed of light is c whether the source is moving or stationary, but the measurement of that speed will not be c if the measurement point is moving. If the speed of incoming light from stars or other sources, in each of the x, y and z axes is measured and found to be c in all cases, then the measurement point is stationary, and that forms the essential absolute reference frame.

Thus, Einstein progressed with his theory whilst the most fundamental fact is missing.
Step 2. His concept of inertial frames would not have been necessary if he had defined an absolute reference frame in step 1, and because he did not define an absolute reference frame, he had to state incorrectly that the laws of physics must apply in every inertial frame.
Step 3. Einstein's theorem of addition of velocities cannot be applied to

light for two reasons;
(1) The decision conflicts with Maxwell's work in which he explains that light-waves are self-propagating because of the interference between the two fields, and this sets the constant speed of light at c, regardless of the velocities of the surrounding objects, or frames. Furthermore, the speed of the source of light, such as the moving train, does not change the speed of light.
(2) An absolute reference frame can be defined so there is no requirement to create inertial frames. There is no requirement to state that the laws of physics, such as the speed of light, must apply in all inertial frames, so that he is wrong to state that the speed of light in a moving train is c, relative to the train. Light travels at c relative to the absolute reference frame of stationary, therefore an observer on the embankment will see the light in the train travel at c, not at c − v, exactly the same as light passing along the embankment.

Thus, his theorem of the addition of velocities simply cannot work for light. Einstein's solution would mean that a ray of light must somehow change speed as it passed through inertial frames that have different velocities, yet Einstein does not explain the mechanism that would allow this to occur.

This assessment means that there is no requirement for Einstein's postulates, because there is no difference in the speed of light between stationary and moving places or sources.

Einstein's decision is not correct. He should have explained in his decision, why he believes light should behave in the same way as an object as he states in his introduction to his theory, when Maxwell had already made it clear how light works.

Step 4. If one accepts the laws of physics to be correct, as one must, and that stationary can be defined, then Einstein's postulates are not required, and are not correct statements.

Step 5. There is no conflict between physics and relativity because Einstein's velocity conclusion in his relativity is incorrect. The motion of the train cannot change the speed of light because the law of physics states that the motion of its source does not change the speed of light.

But it was the different observation of the speed of light in relativity (c − v), and in physics (c) that led him to believe that there was conflict between relativity and the laws of physics, but that conflict does not exist, because when an absolute reference frame is defined, the laws of physics must prevail.

One cannot simply rule out an established law of physics just because it does not agree with his addition of velocities and Einstein's entire theory should have stopped at this point.

Einstein should not have pursued his theory further because by doing so he is claiming that a law of physics is wrong. If he wished to continue, he should have explained why that law is wrong in classical physics terms. As mentioned earlier, the law in question is that explained and mathematically proved by Maxwell, that light is self-propagating and drives itself to travel at the constant velocity, c, regardless of surrounding velocities. It is not possible for any other velocity to influence the speed of light, and that is why Einstein's 'addition of velocities theorem' does not apply to light as it does to an object, but he does not explain why the established law is incorrect, therefore his theory is invalid.

The key point, and the reason Einstein's theory is incorrect is; Because absolute stationary can be defined, which Einstein did not do, the physics laws of light-waves must take precedence over the concept of relativity.

Step 6. Einstein failed to consider the reason that the velocity of an inertial frame cannot be added to a light wave, and failed to explain why Maxwell's established law of physics is incorrect. By pursuing the resolution of his perceived conflict via relativity, a conflict that does not really exist if he had defined the absolute reference frame, he is breaking the laws of physics so that his theory cannot be accepted by scientists to be correct.

Steps 7. Einstein's use of mathematics to prove his postulates to be correct, and to remove his believed conflict between relativity and physics, was neither correct nor necessary. There is no such conflict if an absolute reference frame is defined and the law of light waves is properly understood.

Einstein proceeded to employ mathematics and claimed to have achieved his objective of proving his postulates correct. So, the remaining question is, how is it possible for his mathematics to show that his 'principles of relativity' are correct if they break a law? The answer is that he continued to apply assumptions of relativity rather than laws of physics, and that action is bound to prove relativity to be correct, and not physics. That discussion follows.

Einstein's mathematics.
The only way that mathematics can support a theory that breaks a law of physics is if the assumptions he has applied also break the laws of physics, and that is what Einstein has done.

Einstein's equations are shown only as an appendix because they are not the cause of his error. His equations show that his postulates are correct, but that is only because he included his 'principles of relativity' assumptions rather than assumptions that are laws of physics, therefore one can say his mathematics is not valid because the laws of physics must always take precedence over unestablished theories. The assumptions he used, in his own words, are as follows.

1. *"When we take into consideration the fact that light in the moving system when viewed from the stationary system, is always propagated along those axes with the velocity $\sqrt{c^2 - v^2}$"*
2. *"Light, when measured in the moving system, (relative to the inertial frame) is always propagated with the constant velocity c (as the principle of constancy of light velocity in conjunction with the principle of relativity requires)".*

Overall Assessment of Einstein's theory of Special Relativity.
If Einstein had realised that an absolute reference frame could be defined, he may not have produced his theory of Special Relativity at all, but because he did not realise that, and instead introduced the concept of inertial frames, he introduced a form or relativity between frames, where nothing can be defined as constant, and so he introduced a level of complexity that was not necessary. Physics and nature are simple, and when such complexity is felt to be necessary, one must consider that there must be something wrong with the approach to the problem.

Einstein was unable to define an absolute reference frame so that his theorem of addition of velocities had to accept that the train and the embankment were simply inertial frames that moved relatively to each other at velocity v. His statement that the speed of light seen by an observer on the embankment is less than c is not correct. When stationary is defined and the laws of physics are applied the speed of light seen by the stationary observer is c, the same as light passing along the embankment.

Einstein could have created a 'theoretical reference frame' by simply deciding to call the embankment 'stationary' and then the laws of physics would mean that the speed of light in the train, seen from the embankment is c, but perhaps that would not have been seen as an acceptable solution.

However, the absolute reference frame of stationary is easily defined, and having done that, the laws of physics must prevail over Einstein's concept of inertial frames, and that is Einstein's error.

Failure to define stationary, and his incorrect decision about the speed of light, led to his need to devise his postulates, separating 'relativity' from 'the laws of physics'. But when stationary is absolutely defined, there is no requirement for his postulates. There is no cause for the conflict that he believed existed between the two, and the postulates are wrong in terms of the laws of physics.

This, in turn, means that all of his mathematics that Einstein used to prove his postulates to be correct, was redundant. It achieved its goal only because all of the assumptions used were not those in accordance with the laws of physics, they were relativity assumptions. Thus, his mathematical proof is false.

Einstein's decision to use Lorentz Transformation [5] was inappropriate because all of his logic and decisions up to that point were incorrect. The theory of relativity can never take precedence over the laws of physics, unless he also demonstrates that the laws of physics are wrong, and he did not do so. Therefore, his equations and final conclusions about length and time have no basis in physics. They must remain as 'visual perceptions' as was the basis on which Lorentz produced his excellent work on transformation.

Einstein should have accepted that the laws of physics must always be met and should have used them to determine how to define 'stationary'. The laws of physics are tested and written to describe the real world in scientific terms. If a theory does not reflect those proven laws, it does not reflect the real world and cannot be correct.

Einstein's theory of Special Relativity has no basis in physics, and this means his General Relativity theory also has no basis in physics. Einstein has not shown that gravity is caused by curved space.

Einstein's theories of Special Relativity or General Relativity cannot be retained simply because they appear to provide solutions to phenomenon observed in the universe, if an alternative theory of gravity can do exactly the same. A separate submission will propose an alternative theory of gravity that satisfies all of the same observations that are said to prove Einstein's theory is correct, such as light lensing and Mercury's orbit.

Conclusions.
1. The flaw in Einstein's theory of Special Relativity is his failure to define an absolute reference frame.
2. Without applying any physical argument, he constructed inertial frames using logic that goes completely against the laws of physics. He could have defined his own simple version of stationary in order to clarify the situation in terms of physics, but did not.
3. He incorrectly ruled that his concept of inertial frames should take preference over physics laws, and that decision was a fundamental error, and cannot be accepted by any scientist.
4. It must be accepted that, because of that incorrect decision, Einstein's theory of Special Relativity does not have a sound scientific basis, and so is invalid.
5. This means that his General Theory of relativity is incorrect because it relies upon his conclusion in Special Relativity that length contracts at increasing velocity, which, from the above discussion, has no basis.
6. Therefore, there is no basis on which to state that curved space is the source of gravity.

The following is Einstein's mathematical proof that his postulates are correct which I am including just for completeness, but you can skip the next few pages as they do not contain any more errors.

Einstein's use of incorrect assumptions in his mathematics.
It is not necessary to repeat all of the equations in Einstein's paper but merely to explain and demonstrate his errors.

Einstein applied his relativity assumptions in his mathematics in an unbalanced way just as he did in his logical approach in his 'Addition of Velocities theorem' as described earlier.

Looking now at the detail of his mathematics,

His long equation expressing the coordinates in the moving system is correct, and he then seeks to transform those coordinates to the stationary system.

$$\frac{1}{2}\left[\tau(0, 0, 0, t) + \tau\left(0, 0, 0, \left\{t + \frac{x'}{c-v} + \frac{x'}{c+v}\right\}\right)\right]$$
$$= \tau\left(x', 0, 0, t + \frac{x'}{c-v}\right)$$

He uses the assumption,

"When we take into consideration the fact that light in the moving system when viewed from the stationary system, is always propagated along those axes with the velocity $\sqrt{c^2 - v^2}$"

This relativity assumption follows naturally from his decision to pursue a relativity approach in the mathematics of removing the conflict between relativity and the laws of physics that he believed existed. The assumption is one of relativity and not one of the laws of physics.

If he had applied the correct assumption of c, then the subsequent equations would represent physics rather than relativity.

We can now see that,

If his approach does not ultimately show that his postulates are correct, then his choice of using relativity assumption in all of his mathematics, cannot lead to equations and conclusions that can be described as being a new law of physics. They are simply expressions of relativity.

Lorentz Transformation.
Continuing with Einstein's logic in his book **(section 11),** Einstein claims that two classical understandings are unjustifiable.

1) The time-interval between two events is independent of the condition of motion of the body of reference.

(2) The distance between two points of a rigid body is independent of the condition of motion of the body of reference.

He argues that if we drop these understandings, the incompatibility between light and relativity disappear and his addition of velocities becomes invalid. Thus, he raises the question,

'How do we modify the considerations of the 'addition of velocities' in order to remove the conflict between light and relativity"?

This leads Einstein to ask,

"How are we to find the place and time of an event in relation to the train, when we know the place and time of the event with respect to the railway embankment"?

Einstein then stated, it is easy to obtain the magnitudes of ξ, η and ζ if we express by means of equations and assumption that, and he states,

"Light, when measured in the moving system, (relative to the inertial frame) is always propagated with the constant velocity c (as the principle of constancy of light velocity in conjunction with the principle of relativity requires)".

The purpose of his subsequent equations after including this incorrect assumption was to show that his postulates were correct. If we now study his mathematics that follows, we see that the assumption c he chose to use would inevitably produce exactly the same equation as for the stationary frame.

In terms of physics, the correct assumption for the speed of light *measured in the moving system relative to the inertial frame* **is c – v,** where v is the velocity of the moving system.

If Einstein had used that correct classical physics assumption, he would not have proved his postulate to be correct, but by using the wrong assumption c, his postulate is shown to be correct.

Therefore, Einstein has applied a relativity assumption rather than a classical physics assumption, and having done that, his final equations must remain as expressions of relativity and not expressions in new laws of physics.

His solution for ξ can only be an expression of relativity.

Einstein's claim that his postulates are correct.

Einstein discusses sending a spherical wave outward from the common origin of the two systems of co-ordinates with velocity c in the moving system, (that is his assumption for the light speed as above) and derives the following equations

$x^2 + y^2 + z^2 = c^2 t^2$ for the stationary system, and

$\xi^2 + \eta^2 + \zeta^2 = c^2 \tau^2$ for the moving system.

And he states, *"Therefore the wave is propagated in the moving system with the same velocity c"*,

He is saying that because he believes that he has proved that the speed of light is c in both the moving system and the stationary system, he believes his approach of relativity beats the laws of physics and that his postulates are correct. This means that he believes that the two equations produced in Lorentz transformation can be interpreted as being real and not simply expressions of relativity.

But the weakness here occurs firstly from his original train of thought that if stationary cannot be defined, the laws of physics must apply in all of his theoretical inertial frames, and secondly in his use of relativity assumptions.

If instead, we use the classical physics assumption that the speed of light in an inertial frame is c - v, and is not Einstein's c (and this is valid because the laws of physics apply with reference to the absolute frame), then

The equation for the moving system, using the correct physical assumption, changes to

$\xi^2 + \eta^2 + \zeta^2 = c\ (c-v)\ \tau^2$ which of course is not the same as the stationary system.

Because he incorrectly chose to use the relativity term, and his incorrect postulate value, in the above assumption of c rather than the correct physics term c-v, in terms of physics, he has not proved his postulates to be correct.

It is clear that whatever assumption he chose to insert in his equations would define the result of his postulates. Thus, when he applies the relativity assumption above, that the speed of light is c in an inertial frame, that is what his equations will show. Equally, if he had made the assumption that the speed of light is c -v in an inertial frame, that is what his mathematics will show.

Because of his incorrect use of the assumptions, he failed to prove that his postulates are correct.

The implication in his final mathematical equations is that because the assumptions he inserted are relativity terms, it means that his conclusions are also relativity terms, so that he has really simply achieved the result one would expect when using Lorentz transformation. There is no error in his equations. They simply describe how a stationary observer will see a moving rod and a moving clock

Because Einstein believes that he has shown his postulates to be correct, and he has done so using relativity, he believes it reasonable to state that the equations produced by Lorentz apply to inertial frames and his postulates, and these are shown below. But that assumption is not correct, and the two equations actually express how a stationary observer sees time and length in an inertial frame.

The two Lorentz equations are,

$$\left.\begin{array}{l} x' = \dfrac{x - vt}{\sqrt{1 - \dfrac{v^2}{c^2}}} \\[2em] t' = \dfrac{t - \dfrac{v}{c^2} x}{\sqrt{1 - \dfrac{v^2}{c^2}}} \end{array}\right\}$$

Einstein's theory cannot be retained simply because it appears to provide solutions to phenomenon observed in the universe if an alternative theory can do exactly the same, and that is the case as I discuss towards the end of this chapter, and in chapter four.

Lorentz did some brilliant work to show mathematically the way to calculate how a stationary observer will see a moving object, but I have not included either Lorentz, or Einstein's detailed mathematics, as I do not want to infringe copyrights but you can find Lorentz Transformation, and Einstein's paper and his book, 'Relativity: The Special and General Theory', on the web, and Einstein's Lorentz Transformation mathematics is appendix one. But as I am arguing that Einstein's maths is unnecessary nonsense there is not much point in showing it.

Mathematics is a virtual subject, and I discuss this near the end of this book. Numbers and algebra are meaningless unless accompanied by facts. A well-known example of this is that an equation shows that there are 11 solutions and that is interpreted to mean that there are 11 physical dimensions. Some scientists argue that the extra 8 dimensions must be curled up so tightly that they have no effect, but we know that there are only 3 dimensions, so the extra 8 dimensions are virtual solutions. They are impossible in reality. (I am not sure that a mathematician is able to understand the practical aspects of physics, but physics must be viewed as a practical subject or it has no value).

Similarly, if mathematics is to show that a solid metal rod will reduce in length at very high velocity, the maths must be based on the energy and forces that physicists agree can shorten the length. If the maths is based on 'visual perception,' and the constancy of the speed of light, it is meaningless in terms of physics and 'reality'.

So, the problem is that Einstein's maths is numerically and algebraically correct, but it is based on an incorrect logical argument and does not contain a jot of physics, it is a virtual solution based on 'visual perception' maths and it goes against the solid work of Maxwell. But because of Einstein's logic errors, his maths is not required at all

This is a serious problem. It explains why these ridiculous theories have survived for over 100 years, but what is worse – it is impossible to *prove* that the theories are nonsense, it can only be done by arguing that his theories were formulated on wrong ideas and assumptions.

I would stress that solving the errors of Einstein was not easy. His theories have stood the test for over 100 years. Why has it not already been solved?

But 'I am to go boldly (and un- grammatically) where no-one has gone before' to declare that the theories are total nonsense, and argue that there is a better theory of gravity than 'curved space'. That is really the purpose of this book. If you choose not to believe my argument, and to not scrap Einstein's theories, then another 100 years will pass before gravity is understood.

Examine the sketches used to explain time dilation.
It is worth looking at the sketch in the website 'Special Relativity for Dummies' because it is easy to find on the web, simple to explain, and total nonsense. It was this sketch that gave me the opinion years ago that Einstein's theories were wrong.

There are many similar sketches and videos on websites and in books that explain Special Relativity in simple terms. They may show a spaceship travelling horizontally with a light pulse passing through the length of the spaceship and being reflected back by a mirror, or the light pulse may pass vertically across the spaceship and be reflected by a mirror, or it may be a train with a light pulse sent out from the front, but all of them contain the same error, in that they fail to correctly show how the moving observer sees the light pulse. If you were to accept Einstein's postulates, that would be the correct description, but it is not.

The sketches are all drawn in a way that complies with Einstein's mathematics, but if you study them carefully, and apply a little common sense, you will realise that what is drawn is impossible because they do not show correctly, how the light pulse is seen by the moving observer and that means that the length of the light pulse seen by the moving observer is not drawn correctly when it is compared to the length of the light pulse seen by the stationary observer. The reason for that is that the velocity and distance travelled by the moving observer is not included.

For example, in the sketch of 'SR for Dummies', where a spaceship is moving fast horizontally, the real situation is that the astronaut will know that he is moving because his feet will appear to him to be several centimetres behind their actual position because his eyes will have moved forward before the image of his feet reaches them. A vertical light pulse has no lateral momentum so that it will be left behind as the spaceship moves horizontally to the right and will miss the mirror completely. The astronaut will see the actual light pulse pass backwards and diagonally downwards.

When the motion of the observer is correctly shown, there are two events; the stationary observer seeing the light pulse, and the moving observer seeing the light pulse. When the equation time = distance / velocity is applied *separately* to each of these events, there is no time dilation.

In the early days of this project I assumed that these sketches correctly represented the mathematical conclusions reached by Einstein and that he must be right, but the sketches are nonsense and they show that Einstein's theories are nonsense.

There is an important decision for you to reach at this point.

"Are you going to believe in Einstein's theory because everyone else does, and they even say that he was a genius, in which case you will decide that the sketches, and all the statements made by Einstein in his paper must be correct and you are perhaps failing to understand them. Or are you going to consider the real physics of the situation, using common sense and the laws of physics that you were taught in school, and decide that Einstein was not a genius, he made simple mistakes and the theories are wrong? And if Special Relativity is wrong then General

Relativity, curved space gravity and 'space-time' are also wrong."
In my assessment Einstein was no genius. The fact that quantum solutions for Einstein's gravity and his other conclusions, have not been found in over 100 years is sufficient proof that something is wrong. But there are many websites that try to help you to understand his theories, and that means such writers all believe the theories are correct. I have not found a single website that argues the theories are wrong. That is a major problem. How-ever did physics get into this situation?

It is clear to me that none of the sketches describe the real situation, and they actually show that Einstein was wrong. All that Einstein really had to do was to find a way of defining 'stationary'.

Now examine the pure physics rather than the mathematics.
There is no physical proof that metal rods change length or that clocks go slow at high velocities.
1. Length.
If we looked at Einstein's conclusions in pure physical terms, Mathematics can produce only a virtual solution unless it describes, and is formulated, on the physical means by which the result is achieved.

Where there is no change in physical state, such as when simply moving a distance between two points, it is just a change in position, and that can be changed by velocity and time. One can state that 'distance' is achieved by the equation distance = 'velocity / time', and that is a law of physics that cannot change.

But changing a length of an object is a change in its state. The mathematics must be formulated around the mechanism by which that change in state is achieved, and not simply by velocity and time, and in most cases that would imply energy and forces. If the chosen style of mathematics cannot cause a change, then the change is virtual and does not really occur.

One cannot conclude that a metal rod will become shorter at high velocity if the force to achieve that reduction is not included within the mathematics. Tremendous force is required to compact the length of a rod, but less force would be required to shorten the length of a rope. How are these different forces explained? How can an equation based on light that appears to prove that a metal rod reduces in length also apply to a length of rope?

The *'length' of an object* is not the same as *distance* in the above equation. Length is a measurement of an object using a known length. There is no physical evidence that a metal rod reduces in length at high velocity, and it would require a force that is so far not defined to produce it, and not an equation based simply on velocity and light. So, the conclusions that time dilates and length contracts are virtual and not real because the conclusions are based solely on the mathematics of velocity and time.

2. Time.

It is said that experiments of a clock in an aeroplane proves that time goes slower at a high velocity. In fact, the clock will go slower, but that is because of physics not Einstein's mathematics. The energy added to the plane to produce velocity adds mass to the clock, and that makes the mechanism of the clock harder to work. Extra energy must be added to the clock if it is not to lose time. Any idea that a spaceship travelling through the universe somehow has a slower time, is nonsense.

Time is just a mathematical expression and takes no physical form. One could argue that time is just an invention to enable us to define velocity and compare velocities. The equation Time = Distance / Velocity is the definition of time, and the definition remains correct whether velocity is the maximum of c, or any lower velocity. Therefore, time cannot slow at high velocity. Time has no 'speed'. To slow time would require velocity to slow, or distance to increase, or the equation would be invalid, but the equation is a law of physics.

We have devised a measurement of time based on the event of one revolution of our planet Earth to be 24 hours and that is a fixed time that does not vary because it is a fixed event. We live within that 24 hour event that repeats itself every day so it appears that time actually exists, but it is the event that exists and we simply measure it in mathematical terms.

Energy causes changes to occur and thus velocity, so the image that we see as 'now' is constantly changing, but energy alone is required to achieve these changes. But all the uses of energy that occur produce separate events, and so the time that each generates is separate. The times cannot all be joined together to form a continuum. We may perceive them to be linked together, but they are not. The time for a bird to fly across a garden has no relationship to the time a plane flies overhead, or the time for a car to pass by. They are all separate independent activities.

'Space-time' is a rather grandiose title and it is purely a relative term. It is not possible for time to pass at a different speed depending on where you are, or how fast you are travelling.

So, the story about the two twins is nonsense and I will just explain the believed paradox.

In relativity, stationary does not exist and all motion is relative, so the twin on Earth is travelling at the same relative velocity as the twin in the spaceship. But the travelling twin has to accelerate, then stop, then accelerate back to Earth, and this all takes time that the stationary twin does not need, so when the twin returns to the spot where the stationary twin is, the travelling twin has aged more than the stationary twin. None of that is correct because stationary can be defined.

Motion must slow at high velocities due to increasing mass, so the twin in the spaceship will have a watch that goes slow, a heart-beat that is slow, his whole metabolism will slow down and he may die. But he will be no younger than the stationary twin. Both twins will have experienced the same number of years and are the same age.

So sadly, I have to conclude that Einstein's Special Relativity, and all of its conclusions about space and time, are nonsense, and that means that Einstein's theory of General Relativity and curved space are also nonsense.

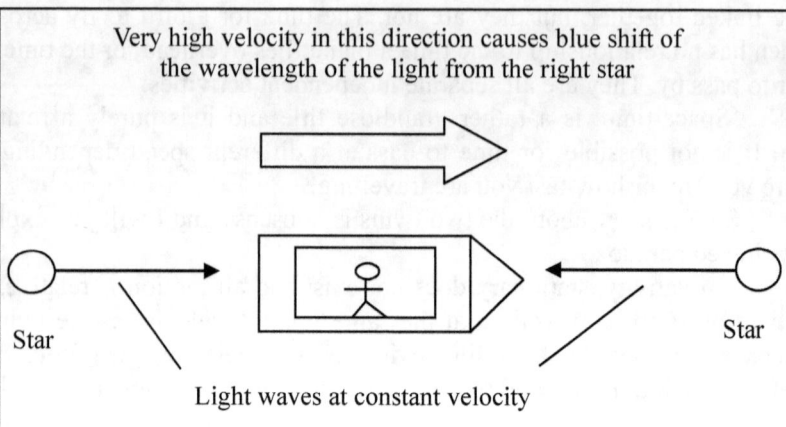

Very high velocity in this direction causes blue shift of the wavelength of the light from the right star

Star   Star

Light waves at constant velocity

Special Relativity is based on Galileo's assumption that it is not possible to define the state of 'stationary'. All objects simply move relatively to each other. The sketch shows that all velocities can be compared to the speed of light either by measuring the velocity of incoming light waves in the x, y and z planes. If these are exactly c, you are stationary. Or by measuring red and blue shift. Light of identical wavelength reaches the spaceship from two stars, one each side of the ship. If the wavelengths are the same, the spaceship is stationary. If the astronaut sees blue shift from the right star, the space ship is moving towards the right star, and the left star is seen with red shift. This can be repeated in three dimensions. Motion must be compared to the speed of light. So 'stationary' is a definable state.

Galileo's incorrect assumption that 'stationary' cannot be defined, formed the basis of Einstein's Special Relativity

Einstein's General Relativity. (GR).

Einstein's theory of gravity, which is widely accepted, is based on the theory that length contracts with velocity, but there are considerable questions on whether rotation can form the basis of gravity when Newton said the basis is mass, and it is doubtful whether the curved fabric of space would cause objects with mass to curve, even if curvature was a correct theory. And do all massive objects rotate?

I am sure few people care whether length contracts or not, but it is the basis of Einstein's gravity, and therefore it matters. I will explain how Einstein came up with the idea that space is curved, and that curvature causes gravity.

My sources of information were Einstein's book, 'Relativity: The Special and General Theory', and the Wikipedia explanation of GR.

In my opinion, in addition to my earlier conclusions on Special Relativity, the entire theory of GR can be dismissed with a few statements, as follows,

- The continuing force of gravity requires energy to sustain it, but Einstein does not mention energy. A force must have a quantum explanation to be valid, and GR does not.
- A solution based solely on geometry is not possible. The universe has dimension and therefore must be filled with material. That can only be electromagnetic waves. Even if his argument is correct that these waves are curved, objects with momentum passing by would continue straight and pass through the curved waves.
- Neither Newton nor Einstein explain why gravity causes acceleration, and one could add, that they do not explain why Earth rotates.

But Einstein has made everything extremely complex with his use of fictitious forces and complex geometry.

I will cover some obvious errors first.

Gravitational red-shift

It is known that light waves in a gravitational field do change frequency and that is because gravity is a force and has energy that acts on the light wave. I explain this in the later chapter on gravity.

Einstein's explanation is to consider a rocket spaceship that is accelerating upwards far away from Earth's gravity, in which there are two observers, one high in the rocket and one lower. When the lower observer sends a light up to the higher observer the velocity of the observer, relative to that of light, causes the light to *appear* to be red-shifted. Similarly a light sent down from the higher observer to the lower observer *appears* to be blue-shifted.

Note that these are just optical effects, not real energy changes, and occur only because of the relative velocities, they have nothing to do with acceleration or energy.

Einstein then said that such frequency changes must occur within a gravitational field, because that is acceleration. But relative velocity and acceleration are not the same. Furthermore, what the observers see are just optical effects caused by relative motion. In his explanation, light does not actually change frequency. A stationary observer on Earth would not see or measure any change in the light wave.

So these two things are completely different. One is an optical effect caused by relative motion where no change to the light actually occurs. The other is an actual change in the frequency of the light because of the effect of the gravitational force on the light wave.

Thus, so far, Einstein's logic is faulty, and he is assuming that the acceleration of gravity and acceleration of spaceship are the same when they are not. Gravity is a force that pulls light. The spaceship has no energy to affect light, it is causes just relative motion. So his error is comparing energy effect with motion effect and getting it wrong. There is absolutely no point in comparing real change with perceived change.

Time dilation in GR

This concept is incorrect as discussed in SR, but Einstein repeats it in GR. He proposes that time must have changed speed in the above example because the higher observer sees the wave to have a lower frequency than that seen by the lower observer, thus time must be passing at a different speeds for the higher and lower observers, and that also must occur in a gravitational field.

But this is now total nonsense. The frequency changes seen by the observers are just optical effects. The light has not gained or lost energy, nor is it travelling faster or slower. It has not changed in any physical way. A light wave in a spaceship cannot be changed just because the spaceship is moving or accelerating. There is not even any interaction between the light and the spaceship. It is not the same as the 'perception' by people in an enclosed lift. Maxwell and Michelson Morley showed that light travels totally independently from its surroundings at a constant speed and is self-propagating.

The frequency changes have nothing to do with acceleration or velocity. And to suggest that time is passing slower is nonsense because *nothing has actually happened.* It is just visual relativity by observers due to velocity differences.

It is also his error to claim that the time on clocks with each observer shows time is slower for the lower observer than for the higher observer. In this specific situation there can be no difference in the time shown on the clocks.

This is where his relativity fails because Einstein is trying to compare real physical changes with optical perceptions or observations where no physics changes actually occur.

Einstein's solution for gravity

Einstein discusses the 'fictitious forces' arising from acceleration that he argues to be relevant, such as when a disc is spun rapidly, anything resting on the disc will fly off by the centrifugal force.

But this will consume energy from the disc that must be replaced, but he does not say from where. However Einstein decided that the constant pull of gravity is the same as a fictitious force.

He describes that if rods and clocks are on the rotating disc placed at the centre and at the rim, the clocks would go slower on the rim and the rod would be shorter because of the higher velocity, and based on his SR conclusions, but this is nonsense.

He then states that if the shorter measuring rod is used to measure the circumference and the diameter of the disc and divide the numbers the answer is not 3.142, and from this he concludes that normal geometry rules no longer apply to a rotating disc.

The rotating disc that produces a centrifugal force outward can be re-interpreted as a gravitational force pulling outward, because of his SR conclusions of no absolute frame of reference. Thus, he concludes that the time dilation and rod shortening must apply in a gravitational field and that space coordinates x, y and z can no longer be accurately defined.

So, he has identified 'fictitious forces' that occur as a result of acceleration, but without mentioning the word 'energy', and says that this is analogous to a change from straight line coordinates to curved coordinates.

He explains these fictitious forces by defining a rotating reference frame. (The disc). Then he says that these are called tidal forces because they are the same forces that cause tides in the sea.

So I think he is suggesting that if the spinning disc is replaced by the rotating Earth, the space at the rim must be shorter than near the centre and therefore must be curved, and he clarifies that by changing the x, y and z coordinates and applying maths to arrive at a set of equations. But it is all wrong because length does not contract with velocity.

So basically, the error in GR is that length does not contract at higher velocities, and he has confused optical effects caused by relative motion, with actual changes in physics, and his logic completely ignores the need for energy.

But all motion requires energy. If he means that the tides get their energy from the rotation of the Earth within the gravitational field of the moon, the energy of the tidal force must come from the energy of rotation of Earth, and that means the rotation of Earth must gradually slow down! I

don't believe it is slowing. The rotation has always been once every 24 hours. This is back to front because I show in the chapter on gravity that the **Earth is caused to rotate by gravity** and that is why the speed of rotation never changes.
So, Einstein's concept of a tidal force that doesn't require energy cannot be right.
He continues to explain how geometry of space is affected by mass but as he has made so many errors so far I do not see the point of discussing his complex argument about curvature and tidal forces.
So, one error is that, from SR, he says things get shorter at higher velocities – they don't, and that last bit summarises the whole problem of GR. No energy is mentioned anywhere in his theory, yet things are caused to move, and that is not possible. It also explains why a quantum solution for GR is missing. If Einstein proposes that motion is possible without energy, there is nothing in the theory to which a quantum solution can be applied. Many later scientists have used quantum field theory, ground state fluctuations and zero-point energy to find a quantum solution for Einstein's gravity, and failed, because there isn't one.
The fundamental description of the theory, as said by others, is that 'Matter tells space how to bend and space tells matter how to move'. The concept is explained that when a spaceship accelerates through flat space, the geometry of space would 'appear' to be curved.

Light lensing

The last problem is that 'light lensing' is believed to be proof that Einstein's curved space is correct. It is a completely logical solution. There is no doubt that light does bend around a massive object such as the sun, and the curvature can be photographed when there is a solar eclipse, but that does not mean Einstein is correct.
Gravitons - Positive radiation from protons in planet Earth, will attract negative light waves passing by and cause them to curve as they pass. That is a practical physics solution, not a theory based on very doubtful mathematics

GR simply raises more and more question

I read an explanation that said "An apple falls because it falls into a deep hollow in space-time caused by the mass of the Earth" Now that is simply 'Alice in Wonderland' stuff! What defines where this deep hollow is? It would need to be everywhere all at the same time?

I can understand the two-dimensional drawings where space is like a trampoline and is bent by the Earth so that the passing moon, which of course has velocity, travels around the bend, but beyond that I have a problem! I cannot visualise what happens in three dimension or when the object has no velocity, and I do not see how it can make an apple fall from a tree, as explained by Newton?

GR does not really help to explain Newton's gravity. What is needed is some practical physics that explains Newton's gravity, and that is what I have done in chapter four, and summarise here.

The real question is - if Newton says gravity is a force acting at a distance and is directly related to their masses, how can there be acceleration? And the answer is that gravity is not a force acting at a distance, it is radiation, and an object moving towards such radiation will receive blue-shifted energy, and so will accelerate. Simple! Why invent bent space?

And there is proof that space is not curved

Experiments in the USA, accurately measuring the position of two stars relative to Earth and forming a triangle, have shown that space is not curved because the angles add up exactly to 180 degrees, and that would not be the case if space were curved.

And I have a similar problem when I look up at the stars, or at a sunset. Nothing is curved. The sun rises and climbs exactly as you would expect it to. The stars rotate across the night sky with no change in their relative positions. How would this be possible if light is curved by warps in the trampoline of space? It is true that light waves from a distant star bend around the massive objects they pass, but there is a simple explanation for this that I discuss later in the book. It does not prove that space is curved by massive objects.

So time does not slow. Length does not contract. The rods on the circumference of Einstein's spinning disc are the same length as those in the centre, nothing is caused to curve by a rotating planet, so space is not curved and it is not the solution for gravity.

Why that has not been realised in 100 years is a matter of concern.

One cannot state that curved space must be right because it explains light lensing and Mercury's orbit, if the theory is nonsense.

I have explained what I believe gravity is in chapter four, and my solution satisfies all the same observations and experiments that are used to claim GR is correct.

The statement that "Acceleration and gravity can be equivalent only if a massive object curves space-time" is simply not true.

There is one more equation that Einstein produced in his theory of Special Relativity and that concerns mass, which is the subject of the next chapter.

His equation is,

Kinetic energy = $\dfrac{mc^2}{\sqrt{1-\dfrac{v^2}{c^2}}}$

The suggestion is that mass will rise to infinity at the maximum velocity as this equation, and Einstein's argument in his SR paper, indicates. But, as discussed earlier, the equation is based on incorrect assumption applied to Lorentz Transformation. It is not correct that velocity can never quite achieve c no matter how much energy is applied.

The processes of physics must take preference over the theories of mathematics, because mathematical theories can be no more than an indication, and may be an error, if not based on, or supported directly by, actual physical processes.

The Lorentz methodology does not include energy or forces, and neither did Einstein's version so that his conclusions about mass are incorrect. But it is essential to understand what mass is, and how it works, before commenting any further on Einstein's opinion, and I do this in the next chapter. There is nothing further to add except to conclude that there is no justification and no requirement for Einstein's theory of Special Relativity,

It would seem that the excitement of Einstein's discovery that time dilates and length contracts, giving him the status of 'genius' has drowned the logic and common sense of the situation, so that the whole matter has become accepted. Space-time, time dilation and Length contraction are recognised as fact, so General Relativity, which defines that curved space is the cause of gravity, has also become totally accepted as fact

I have a serious problem with all of this. I can understand Einstein making some physical errors a hundred years ago when the behaviour of light was only just being clarified, but I do not understand how scientists have not corrected Einstein's theories in the last hundred years.

Chapter Three

THE MISUNDERSTANDING OF MASS AND MOMENTUM

This is the next misunderstanding in physics. Peter Higgs collected the Nobel Prize for his work on the cause of mass, but unfortunately his theory is incorrect as I will show.

The chapter provides a better understanding of resistance to motion, mass and momentum because this is necessary before the following chapter on gravity, and duality at the end of this chapter.

In the previous chapter I argued that Einstein's relativity is incorrect, and so his equation for kinetic energy is wrong from which he concluded that mass will rise to infinity at the maximum velocity. It is not correct that velocity can never quite achieve c no matter how much energy is applied. So I will clarify what is the cause of mass, and this is necessary in order to understand gravity.

Einstein said gravity was based on curved space, which was a function of mass, but I suggested in the previous chapter that this is not a realistic theory, and objects with momentum would not follow the curved field that must exist in space to produce the dimension.

Newton said gravity was a force due to mass, acting at a distance. No-one has yet identified the source of the force, but clearly it must be an electromagnetic wave. To clarify that, it is necessary to look at mass and momentum of charged particles and photons.

The Definition of Mass.

I think confusion begins with the definition of mass. The common definition of mass is - 'That which causes resistance to motion'. But what 'That' is does not seem to have been defined, but it is generally accepted that when a charged particle passes through the Higgs Field, the thick field causes resistance to its motion, and therefore is giving the particle mass.

But mass is also known as something that produces momentum and kinetic energy, so how can mass do both of these apparently opposing things? That is where the Higgs theory starts to become unstuck.

Mass cannot be simply a mathematical number, it must be 'stuff', and the only thing that happens when energy is added to a charged particle is that the field extends and becomes a wave.

Weaknesses of the Higgs Theory.

I have heard it said that the Higgs Field sticks to a particle like treacle to give it mass! That does not sound very scientific!

Mass is defined as that which causes resistance to motion, and the Higgs Field is believed to exist and produce that resistance. But for a charged particle to have velocity and move in a Higgs Field it must already have mass, as energy is equivalent to mass. For the Higgs Field to cause mass there would need to be a clearly defined electrical process that cannot vary, but there is not.

The mass of a particle is a very precise number. The spreading radiation from a Higgs Boson cannot possibly produce the uniform density of field that would be required for such precision.

The velocity of charged particles moving in the Higgs Field would vary and that may affect the level of inherent mass. If a particle later moves into a stronger field than Higgs (if such exists), the mass would change.

When a charged particle is given mass something in the particle must change but there is no definition on how the added mass is stored. Mass cannot be just a number, so what form does mass take?

It is not clearly defined how the mass that is created by resistance can become the mass that forms momentum

There is no proof that the Higgs Field even exists. It is a theory supported by the finding of the boson that could create such a field.

It is believed that an electron would travel at the speed of light if it had no mass. But an electron would not move at all until it was given energy, and if given energy it is given mass.

.   *These problems are sufficient to question whether the Higgs Field theory is right and whether the cause of resistance, and energy-mass equivalence are correctly understood*

A better solution

Combining the accepted work of Maxwell, Planck and Einstein, solves all of these weaknesses.

The mechanism of an electromagnetic wave of a photon as defined by Maxwell must also apply to the wave produced by a moving charged particle. It is not possible for two identical waves to have different properties. A photon comes from an electron and cannot be different.

A new definition of the cause of resistance to motion

There is a complete misunderstanding of the cause of mass and resistance to motion.

Rather than thinking that there must be an external force resisting the motion of a particle, one can think of the resistance as being within the particle itself, through a process of converting new energy into velocity and momentum so that Maxwell's self-propagating wave can provide motion.

So the delay in starting motion is simply caused by an electrical process.

Applying Maxwell's work, the wave produced by a charged particle governs the velocity of the particle, and the wave 'carries' the particle. The 'resistance to motion' is within the charged particle itself, and is the delay time in the electrical process of converting new energy into a longer wave with momentum, that then allows motion. The mass and momentum of a charged particle is all within its wave.

Clearly the energy added must produce an increased field size or length, and as that is a precise electrical process, it must take time.

The concept that that there can be two causes such that some mass is inherent and fixed, whilst other mass is variable with velocity, cannot be a complete solution. There must be a direct relationship between those two definitions of mass, and the Higgs Field theory does not make that connection. Why should one lead to momentum while the other does not?

Resistance is not a once-off event, it exists every time new energy is added, and once an object has overcome that resistance, and achieved a new velocity, the mass takes the form of momentum and the velocity will continue forever. So resistance (mass) is more to do with the time when energy is being added to a charged particle.

The Higgs field cannot produce mass that increases with velocity if its role is defined as that which produces inherent mass.

Thus one can say that the apparent resistance is caused by the process of adding new energy to a charged particle which requires a short processing delay to build a complete wavelength. Then further energy and delay builds a larger wavelength that has momentum. It is only when that first wavelength is completed that it becomes possible for the charged particle to apply its momentum and begin velocity. Then each addition of further energy increases the wavelength after a further process delay.

But this resistance caused by the delay in wave completion is not mass. It is the field that is the mass.

A separate cause of resistance to be considered is that, whilst the universe must be filled with electromagnetic waves to produce its dimension, these waves will occur in all directions and will be of no value as a usable space dimension by a charged particle such as an electron, moving in a single direction. Instead, the wave produced by the electron is necessary for motion because the space dimension it produces is required for it to be able to move at all. Perhaps that is why an electron produces a field. Thus there will be resistance to motion until the electron wave is completed.

Clarification of rest mass and rest energy

There is no rest (inherent) mass or rest energy when stationary, as currently defined, but such mass and energy does exist as soon as the process delay is completed and the first wavelength is completed. The particle then begins to move. This value of mass is the same as the current rest mass and so there is no requirement to change the terminology, as long as the cause is clear, and that the mass is not 'inherent'. The mass used in Kinetic energy is unchanged and total energy must include this initial energy.

The process of producing the mass before a particle moves is the same process that produces mass as velocity increases. Both produce a length of field material, but the length cannot produce momentum until it is a complete wavelength and Maxwell's self-propagation applies. There is no difference in the creative process between rest mass and variable mass. At low velocities, rest mass is used in kinetic energy as it does not increase with velocity enough to affect calculations.

When measurements are taken to find the mass of a particle, it is a measure or rest mass – the energy to overcome resistance / delay, and start to move.

The wave is the mass and momentum of a charged particle and it 'carries' the particle. Every time extra energy is added there is a delay while it is converted into a longer wave before it accelerates to a new velocity. Mass is a precise number because it is created within a particle in a precise process. Gravitational mass is the same process. (Discussed in next chapter)

Whilst there is no rest mass or rest energy in a charged particle, both of these exist in a *stationary atom* because the electrons orbiting in an atom have a wave that is mass. (The atom is discussed later in this chapter)

Creation of wave length and width.

If one now considers the process of creating a wave, additional energy will increase the length of the field so that momentum will increase from Maxwell's equations, and velocity will increase. The width of the wave must be enough to 'carry' the particle, and will therefore vary.

We can now apply Planck's work on a photon because all electromagnetic waves must be the same. The process of energy conversion into a wave, being electric, must have a constant conversion process time.

When energy is added, velocity increases and eventually the velocity of the wave will be reached when the process time is too long to extend the length of field and produce a longer wave because the wave has moved on at a very high velocity before the particle could complete the process. That is the explanation of 'Planck Length', and it must be the same for all bosons and particles because it is governed by the constant process time. There cannot then be any greater forward momentum, so that the maximum velocity has been reached. The size of the wave, meaning the length and the width, is mass.

I visualise a wave as being made up of strands of field, so that the more strands, the greater the width and the greater the mass.

If further energy is added when velocity is at the maximum such that the field length cannot increase, the mass cannot increase, and the extra energy uses the existing maximum mass of the wave transversely as transverse momentum, and creates frequency. This is where Einstein's energy-mass equivalence meets Planck's energy of a photon.

Planck length (h) is not just the maximum length of any electromagnetic wave, it is also the factor Planck uses to define the energy of a photon, as hf, where f is the frequency. He concluded that the length of a wave is the energy to achieve c, and that basic energy is re-used and multiplied many times by the frequency.

But although each boson or particle must have the same maximum length, as discussed above, the width of the wave will be different. A photon wave may have only ten strands, whereas an electron wave may have a thousand strands, thus the size of an electron wave is larger than a photon wave, and that size is the mass. If there are more strands, there is more mass and more momentum, so in terms of energy, h is not constant for all waves and particles.

There is no reason why an electron wave cannot achieve frequency, but it would require massive energy, and the idea that mass can rise to infinity is not correct. When the field length is at its maximum, mass is at its maximum, and there cannot be any further forward motion, so that any additional energy must be stored as frequency in which the existing mass is employed transversely, and not as extra mass, so that, $E = mc^2$ must be adjusted to include frequency as in a photon.

Photon waves and particle waves are the same. Both have mass.

A photon is thought to be massless because it has no charged particle and so cannot collect mass from the Higgs Field, but I have argued that the Higgs field does not exist and it is not the reason things have mass.

A photon of light must have mass because it has energy and a self-propagating wave of momentum. It is not the process delay, (or resistance) that is mass, it is the wave itself that is the mass. (That is further proof that the Higgs Field theory of mass is incorrect)

The process delay in creating a photon wave is done by the source electron. So a photon has mass but it is minute because there is no particle to be carried. Thus a photon is the smallest packet of energy possible (a quanta), and is probably only a few strands wide.

A photon travels at c, not because it is massless, but because it has more energy than is required to produce the maximum field length. The source electron's emission creates the fixed length of field at c from its redundant energy yet there is energy left over to create the frequencies within the photon wave.

There is no reason why an electron cannot achieve the maximum velocity if it is given sufficient energy.

Re-defined energy-mass equivalence.

This relationship can be clarified by merging the equations of Planck for a photon, and Einstein for a particle, because both waves have exactly the same properties and creation process.

Einstein showed that the energy of a particle (which is all in the wave) is defined by $E = mc^2$. The energy above that required to achieve c is stored as frequency, using the same maximum mass transversely, with no forward momentum. Thus in $E = mc^2$, c is constant and mc^2 is utilised many times by frequency, f. (And m is different for each type of particle because it is defined by the field width x Planck length). (Number of strands x length)

Planck defined the energy of a photon to be $E = hf$, where h is Planck's constant. If f = 1, (no extra frequencies) then clearly, h is the energy to achieve the maximum velocity, and this is re-used many time by frequency', so $$h = mc^2$$

But note that h is not now constant as it depends on the mass, the ize of the wave, but energy –mass equivalence becomes, $E = mc^2f$

For the entire spectrum of energy.

(f is unlikely to exceed one for a charged particle.)

A solution for Duality, and Dark Matter.

This is where my concept that a wave is made up of individual strands becomes useful. Because the wave must 'carry' a charged particle as it is its momentum, a charged particle cannot move without its field, and because the wave is created when the particle has velocity, the electron will be carried in a waveform manner as if it were behaving as a wave itself.

When streams of electrons are passed through two slits their waves will interfere to form diffraction lines. The electrons must remain within these diffraction lines because they are its momentum, and because the electrons have no mass or resistance to motion, and can move freely in the wave, the electrons will form diffraction-like dots on a screen as if they were behaving as a wave.

If one assumes further, that if one electron is fired at two slits, and the electron does not possess the unlikely concept of 'Super-position' and pass through both slits at the same time, then it is only the wave that passes through both slits, by separating the strands of the wave and passing some through each slit. The electron passes through just one slit, and is carried by one part of its split field.

Dark Matter

'Dark Matter' is not matter at all. Only a wave has mass, and the waves in the universe are whatever electromagnetic waves give the universe its dimension. It is known that Dark Matter has no charge.

The atom

The electrodynamics of an atom is complex but the same process applies. The electron is held in orbit by the momentum of the orbital wave and there is no angular momentum as the electron has no mass.

Because the wave is fixed at both ends, and the electron is pulled inwards by the proton, the field length is caused to be shorter, so that frequency occurs at a much lower velocity and energy than c. Planck length no longer applies.

A quantum fall is the removal of one frequency as a photon, and the quantum level stabilises when an arriving photon is absorbed to restore the previous lost frequency. There must always be a whole number of frequencies or completed wavelengths, and this sets the allowable orbits.

Thus energy from a photon increases the velocity of the electron and a new balance of velocity and frequency is produced that causes the wave to expand with a longer length of field and carry the electron one quantum level higher. The amount of energy is in the balance of frequency / wavelength and field length and that defines the allowable orbit.

Wave polarity

To understand gravity in the next chapter one needs to consider wave polarity. In every-day life we only see negative waves (photons) striking negative electrons in atoms, and the result is the receiving atoms move in the direction of the received energy. It is where a snooker ball hits a second snooker ball causing it to move forward. That makes good sense!

But kinetic energy has no role at wave level because the mass is too small. It is the polarity of the wave that decides the direction of a charged particle. Energy is not a physical property, it is an electrical property.

The momentum of a wave acts on a particle by transferring energy, not via a 'push', and an electromagnetic wave is either positive or negative depending on the polarity of its source particle. Photons of light are negative because they come from the negative wave that orbits the nucleus of an atom.

In the next chapter on gravity, the wave creating gravity is positive and it is absorbed by the orbiting electrons. But the electrons are already full of the negative waves, so the positive wave of gravity cannot add to the electron's orbiting wave, it must create an entirely new positive wave, and because a positive wave will attract a negative charged particle, it will pull the particle. That is how gravity begins.

Chapter Four

THE MISUNDERSTANDING OF GRAVITY AND WEIGHT

If curved space-time is not the cause of gravity, what is?

The mechanism of gravity is the transfer of momentum of radiation from protons to electrons.

I have said that Einstein's idea that space is curved is wrong, Newton's gravity is correct, and shows that the force is related to the mass of its source, it seems to act from a distance, it accelerates an object, and it follows an inverse square law. The question is how can the laws of physics produce Newton's gravity, and how can the problem of inaccuracy over long distances be corrected?

Gravity is a force that causes two atoms to be drawn together and the larger the mass of the object, the stronger the force. The mass of an object is simply the number and type of atoms (the number of electrons and protons) that make up the object. So something emerges from atoms that pull other atoms and it must be due to the number of electrons and protons because they are the mass. But the number of electrons always equals the number of protons, and so protons may also be involved in gravity.

Now electrons are negative and protons are positive so there is an obvious possible solution. Positive radiation from one object will attract the negative field or charge of electrons in the other object. But 'attraction' is the wrong word.

The Standard Model suggests that protons radiate various bosons. This is a subject for specialists in particle physicists, as it is unlikely to be a w, z or Higgs boson, and is probably an as yet undiscovered positive boson. It would seem logical that it is the graviton, and it would have mass as discussed in the previous chapter. This would produce a positive wave of momentum leaving every atom in a massive object, and because planets are spherical, it would rise vertically from every point on the surface. In an atom the number of protons always equals the number of electrons and if a force is applied to the electrons, they will drag the nucleus with them.

All electrons are carried by negative waves because all are orbiting the nucleus of atoms. When a negative photon meets a negative electron the energy in the photon is absorbed and adds to the existing negative orbital wave, so increasing the velocity of the wave and raising the quantum level. But that means the electron is already fully occupied by a negative wave so that energy from a positive boson wave cannot add to the orbital wave. Instead the boson wave from a massive object must cause the electron to produce a new positive wave. This vertically rising wave, being of opposite polarity, attracts the electron pulling it downwards. The boson energy then transfers to the electron and consolidates the new downward wave. Thus a rising boson creates downward motion that is gravity. This motion then causes the electron to receive blue-shifted energy producing the acceleration, g, so accelerating the atoms of mass m, downwards with force mg. There is no delay time because the energy of boson waves is always acting on the electrons in a second object, so that the first electron wavelength (rest mass) is always exceeded.

Such bosons waves from protons have extremely low energy. They are impossible to detect because they are radiated by every atom, everywhere. Humans radiate them. The atmosphere radiates them. Any measuring device used to try to detect them will itself radiate them, and so cannot detect them, even though all of space is full of them. They exist, but there is no obvious way we can prove it, and apart from gravity, no effects of the radiation can readily be seen.

So gravity only affects electrons, and gravity is not, as thought, a force that acts from a distance. It is not a force that 'attracts', or 'pulls', (although I have used that expression for simplicity) and it is not a force that is due directly to mass, although the size of mass is relevant. For example it is not strictly the mass of the Earth that somehow holds the moon in orbit.

Gravitational mass

Mathematically there is a definition of 'Gravitational Mass' as being the mass that governs the attractive force between two bodies. This is correct except that it is actually the proton radiation passing between the two objects that creates the attraction. Furthermore, the 'pull' exists even if there is no second object. But the strength of the attractive force does depend on mass because it depends on the number of protons in the two bodies, which in turn equates to the number of electrons, and it is the electrons that define the mass. (If you accept that the orbital wave is mass).

The process producing gravitational mass is the same as that for inertial mass discussed in the previous chapter. The problem is that inertial mass is currently defined as resistance to motion, and that currently makes a common definition with gravitational mass impossible.

The logic must be that there cannot be two different definitions of mass. There cannot be two different processes that create mass. Gravitational mass must have the same definition and cause as inertial mass, and that is the field length of a wave produced by a charged particle.

That logic defines that the mechanism of Newton gravity is the same as photons of light striking orbiting electrons and increasing their field length (mass). The only difference between inertial mass and gravitational mass is the polarity of the electromagnetic waves involved.

The wave in inertial mass is always negative. It may be created by a photon of light being absorbed by an electron and adding to the wave orbiting the nucleus, or by momentum transfer when one snooker ball hits another. The wave is the momentum of an electron and carries the electron (which has no mass), and the electrons cause the entire atom and object to move.

The wave in gravitational mass is positive because it is the radiation from positive bosons released by protons in the atoms of an object, which in turn, pull electrons in a second body, and these pull the entire atom and object.

If free movement of an electron is permitted, the attraction would produce a wave of momentum and motion exactly the same as when a photon is added to an orbiting electron except in the opposite direction.

Gravity is a permanent force acting on electrons where-ever they are located, so the process delay period is already completed. The total energy of all the proton radiation absorbed must be greater than rest energy.

Radiation from a proton would create a length of field on the electron, and motion toward the radiation creating blue-shift and acceleration, so that field length increases, the same as for inertial mass when energy is added.

If movement of an electron is prevented, (weight) there is a transfer of momentum / energy, from the radiation to the electron that is absorbed by the electron, creating a downward force known as weight, but no field length can be produced as the electrons have no velocity.

Gravitational mass is a transfer of momentum energy to produce a theoretical equivalent field length that such momentum would produce if motion were allowed, and the momentum pressure from the radiation continues because there is no break in the stream of trillions of radiated photons. The electrons will try to collect the proton radiation and will continually attempt the transfer by producing 'kicks' of electromotive force.

The ability of a particle to produce these 'kicks' will depend on its charge and also on the density of the proton radiation – such as the distance above planet Earth or the field density on the moon.

So in both cases of fixed, or free electrons, mass is the size of field, (length x width) and it is that size (mass) that defines the pull of gravity (mg) on an electron, the same as it defines the motion of an electron after absorbing a photon of light.

Gravity has the same effect as any force applied to an object. It produces kinetic energy, and a force of mass x acceleration, except that for gravity the acceleration is the constant g, and is a pull in the opposite direction, and the mass is that of the complete atom not just the electrons.

The distance between the two masses has the same effect as for two charged particles, and is due to the reducing density of radiation.

The distance between objects affects only the magnitude of the force and not the process. The process that produces gravitational mass is electrical and constant regardless of factors such as type of charged particle or atom.

The positive radiation has no effect on the positive nucleus so a proton can produce mass but has no weight.

Milikan's method of determining mass was actually measuring weight, which I call 'equivalent mass'.

Radiation from protons is positive and only reacts with negative

electrons, and the number of protons always equals the number of electrons, and together make the mass of an object. The resistance, (the time to convert energy into a complete wavelength), is a function of the number of electrons in the object, and is therefore a function of electrons plus protons - the mass. The acceleration due to gravity will always be the same whatever the type of material. Heavy objects will fall at the same rate as a light objects.

As discussed earlier, the acceleration of gravity is because when an object begins to move the electrons in the object receive 'blue shifted' proton radiation. Energy is collected at a faster rate, so the object accelerates producing more 'blue shifted' radiation, thus the acceleration continues and 'acceleration due to gravity' is achieved.

Correcting Newton's equation for long distances

The problem with Newton's equation is that the force produced is instant. But we know that if the sun were to suddenly disappear, its gravitational force would continue to hold Earth in its orbit for about 8 minutes. So we know that gravity travels at the speed of light, yet Newton's equation does not allow for that.

His equation is

$$F = \frac{g\, m_1 m_2}{r^2}$$ where r is the distance between the masses

If we simply replace r with $c \times t$, that is, speed of light x time
The problem with long distance and large masses is removed

The curvature of gravity

Because proton radiation is vertical from a spherical planet the gravitation force will be curved around the planet and the gravitational effect will be the same as Einstein's curved space. The strength of gravity depends on the density of the proton radiation, and that depends on the number of protons and electrons ie the mass of the object, just as Newton showed. The fact that all stars and planets are spherical indicates that the source of gravity is within the star, and is not due to 'curved space'.

Gravity is a vertical force

Although the proton radiation is emitted in every direction from the ground the wave directions will balance to form a vertical flow of energy and a planet – being spherical – sends the majority straight up, even if they have to penetrate from Australia through the planet to the other side first, thus gravity pulls straight down. However, because the Earth is rotating, the waves emitted have a very small sideways force vector as well as a vertical vector, so that gravity is not strictly vertical.

The pull of the Earth's gravity is constant because the number of protons in the Earth is constant, and the effect of this pull, or the weight of an object, depends on the number of electrons in the object. Note that the Earth's gravity comes from trillions of waves within the planet's core and these waves are invisible. Gravity is unrelated to light and so is constant day and night.

At first I assumed that the gravity from protons at the time of the Big bang would pull electrons from far and wide to create atoms, but I changed that idea and I discuss it further in the chapter on the Big Bang.

Gravity and Quantum Theory

Quantum theory merely suggests that energy is transferred only in packets or 'quanta'. This is true for light and each quanta is a piece of a photon radiated from an electron. But it is also true for gravity because proton radiation has the same structure as a photon. Each wave is a packet of energy. Thus this theory satisfies the need for a Quantum Theory for gravity.

Just to clarify things, my term 'proton radiation' means radiation from protons, it does not mean a stream of proton particles. The radiation is actually created by a boson when it decays, and that boson may be the missing 'graviton' that is identified as a requirement in the standard model.

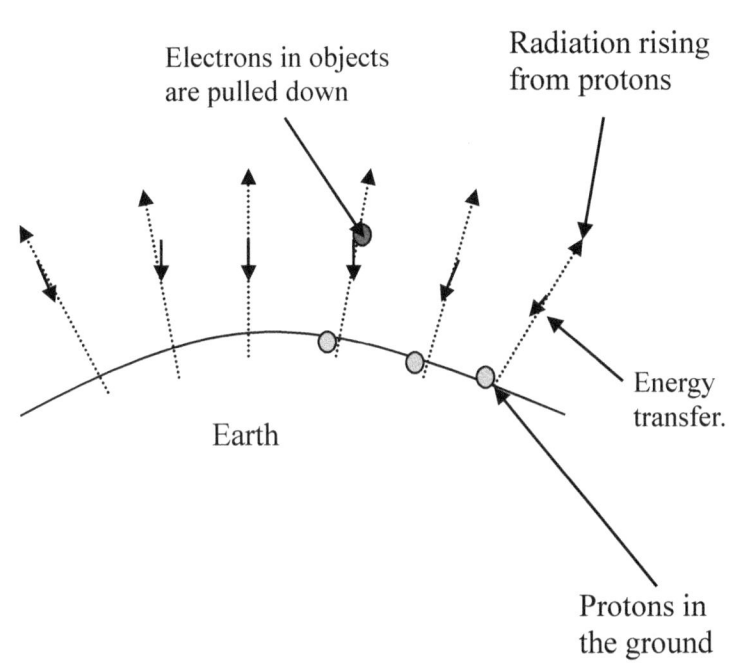

GRAVITY

Trillions of separate waves radiate from protons in the atoms that make up planet Earth. If objects with electrons are free to move, the opposite charge of the proton wave starts to pull the electrons downwards in the opposite direction so starting electron waves. Energy then transfers from the positive waves to expand behind the new electron waves, so producing a large wave that carries the electrons downwards. If electrons are not free to move, kicks of emf give the electrons weight.

Gravitational waves produced by proton radiation

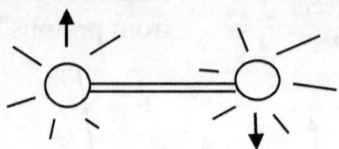

Consider a dumbbell rotating end over end. This represents two massive objects rotating whilst also producing proton radiation. The effect seen by an observer is as below.

When seen in the position above, the observer feels the strength of gravity radiation as,
    1   0   1
When the assembly is rotated so that the observer sees it from each end, the strength is double.
            2
When the two are combined in rotation, it is a wave

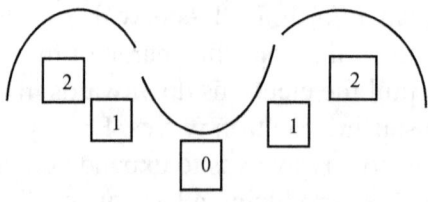

Einstein's field equations, and gravitational waves.

Einstein's field equations are included here only because gravitational waves have been proved to exist, and that discovery is being used as proof that Einstein's equations are correct, but if it is accepted that Einstein's mathematics and physics do not show that space is not curved, then Einstein's field equations have no basis because they are developed from his theory of curved space.

Gravitational waves are described as being 'ripples in space-time'. I really do not know what that means! In my analysis, space is not curved, time does not change depending on where you are, or what speed you are travelling, and in fact later I will explain that time is nothing more than a mathematical calculation for the period of an event and no events are connected to form a time continuum. Space-time simply does not exist.

Gravitational waves are variations in the strength of gravity and those variations will cause light lensing, and red and blue shift. Perhaps that looks like ripples in space-time, but it is not.

The mechanism of gravity that I have proposed can produce exactly the same gravitational waves, and I have described this in the sketch on the previous page using the dumbbell analogy that is normally used to explain it. When two massive objects rotate around each other, their proton radiation will increase and decrease according to their alignment with the observer.

All of the experiments and observations that claim to prove Einstein's GR is correct, can be explained equally well by proton radiation.

I have left the following explanation until now because I needed to explain gravity first. My theory of gravity allows all the experiments supporting Einstein's theories to be satisfied but for different reasons, and so the experiments cannot be said to prove Einstein is right. It just seems to me that scientists are so 'mesmerised' by Einstein's amazing conclusions in relativity that they are attributing almost everything observed in the universe as being proof that his conclusions are right, without even thinking that there may be equally valid solutions that have nothing to do with Einstein.

I also believe that if someone designs an experiment to prove Einstein is right, that is the conclusion they will reach even if there are several other equally valid causes that have nothing to do with Einstein and I think that is what has happened in the following experiments.

Because Einstein's theory does not involve or account for the missing mass in the universe, it cannot be regarded as a complete solution. The theory cannot be demonstrated in a laboratory and so is just a theory. My proposal that gravity is caused by proton radiation is proved by the Casimir Effect, and because the radiation is mass, the universe is filled with mass.

Einstein's theory is so complex and requires such extrapolation of normal physics that it does not seem to fit with the simple laws of physics and common sense. If all of the observations that support Einstein's theories can be explained with a much simpler theory as I have done on the following pages, there becomes little proof that the theory is correct.

But the problem is that so many facts and observations have been attributed to GR that it is now impossible to find them all, and explain them all, and each year more papers are published to make the matter worse. So I am only going to cover the main arguments used to support GR.

a) The Casimir Effect, produced by Hendrik Casimir and Dirk Polder in 1948, showed that two metal plates of high electron content are pushed together, and this was argued to prove that vacuum fluctuation do exist and are able to produce a force or a gravitational effect, but the key phrase is 'high electron content'. Such metal plates will have high proton and electron content. The protons radiate waves that transfer momentum in the opposite direction to the electrons, thus moving the plates together. So the Casimir Effect does not prove the existence of vacuum fluctuations, it actually proves that gravity is a function radiating from an atom that only attracts electrons. *(My theory exactly)*

b) Newton's law of gravity is a simple straight relationship of two masses, but this relationship is unable accurately to define the weird orbit of Mercury. The theory that proton radiation is the cause of gravity satisfies this unexplainable orbit of Mercury.

Proton radiation from the sun onto Mercury will increase the velocity of electrons in atoms during the portion of their orbit towards the sun, and reduce it when moving away, so producing an unbalanced momentum and an attempt to move the centre of orbit. The effect from all atoms will be to cause the entire planet to rotate on an axis perpendicular to the sun's gravity, producing an Equator similar to Earth. This rotation will produce precession along Mercury's orbit path around the sun. The combined effect of rotation and precession are suggested to be the cause of Mercury's unusual orbit. Thus Mercury will travel further around the sun than basic Newton gravity would suggest. This gyroscopic process is explained further in chapter seven.

c) Gravity from a massive object has the ability to bend light waves passing it from a distant star and this is called gravitational light lensing. There are many examples to show that light does indeed bend around large objects, but one cannot conclude that this proves space is curved. Proton waves radiating from every atom in a massive object have an opposite leasing edge polarity to light waves passing the massive object and it is the magnetic attraction of these opposing polarities that causes the light wave to bend. The waves rise vertically from the spherical shape of a star and so the attractive force curves around the star in the same way as 'curved space'.

d) The Shapiro effect of gravitational time delay suggests that light waves passing through a gravitational field take longer than when not in such a field, so time must have slowed. But in my analysis, time is not a dimension. It has no speed' and cannot be slowed. Time is just man's way of comparing velocities, so the suggested time delay cannot be correct.

If planets released positive radiation and this produced the gravitational field, then negative light waves passing through this field would be buffeted and would follow a wobbly and longer path due to the opposite polarity of the two fields, thus the journey of a light wave would be longer in the gravity field. Time does not slow down.

e) General Relativity predicted that gravity causes time dilation and that time dilation produces red-shift and blue-shift of a light wave, but I have just suggested above that gravity does not cause time dilation. The Pound-Rebka experiment proved that the red and blue shifts do actually occur, and I suggest that the real cause of this is the same as light lensing in (c) above.

Proton radiation will attract a light wave that is travelling towards Earth so causing blue-shift, but reduce the energy in a light wave that is travelling away from Earth so causing red-shift, but there cannot be any momentum energy transfer because light does not have a charged particle.

f) One can suggest alternative explanations for Black Holes that have nothing to do with curved space, red-shift or any other aspect of General Relativity. I offer one solution that, if correct, implies that there is no event horizon and that Black Holes are completely misunderstood.

My suggestion that proton radiation is the source of gravity means that intense positive proton radiation from atoms in the centre of the black hole will pull negative electrons down and out of their orbit in atoms so that their energy is released in one burst of gamma radiation. Without orbiting electrons, photons can no longer be produced. It is so simple and uncomplicated that it is probably correct.

g) In March 2014, a team from the Harvard-Smithsonian Centre were able to show a change in the polarisation of light that occurred about 380,000 years after the Big Bang, confirming that the rapid expansion of the universe, known as 'inflation', did occur. Inflation is only possible if gravitational waves exist, so the observation proved the existence of gravitational waves.

i) Muons, being negatively charged particles the same as electrons, would receive the same attraction and momentum energy transfer as electrons described above, carrying them further when moving toward a gravitational field, thus further helping to prove this theory.

I agree with all of those conclusions, but the statement went on to say that the existence of gravitational waves gives the final proof that Einstein's theory of General Relativity is correct, and I do not agree with that because I do not agree with Einstein's theory. Like all of the above experiments, they have again jumped to the wrong conclusion.

I suggest that Gravitational Waves are periods of dense positive radiation such as proton radiation, which is a source of gravity in that they pull electrons, and are produced when one massive object moves in close proximity, or collides with a second massive object.

The idea that space is distorted by vacuum energy into ripples seems illogical to me until the process is much more clearly defined. Ie Space cannot be 'nothing' – an empty void, so what is the material that is caused to ripple, and what is the source of vacuum energy? If the answer to both is proton radiation, then we are in basic agreement, but with minor differences.

My suggestion may not be totally correct, but it is an alternative theory to that of Einstein, which, in my view, cannot be correct.

So the statement that "Acceleration and gravity can be equivalent only if a massive object curves space-time" is simply not true.

The proton radiation is a much simpler way and that in itself is a powerful argument to say Einstein was wrong.

The reason proton radiation theory beats Einstein's GR

It explains why stars and planets spin.
It explains why stars and planets are spherical
It is a quantum solution if the right boson can be found
It offers a solution for dark energy and dark matter
It offers a reason why galaxies are moving further apart
It explains why inertial mass and gravitational mass are the same
It offers a solution for Newton's gravity.

Comparison with the principles of Newton.

Newton showed that the force of gravity is due to the mass of an object and his law of gravity is a simple relationship between the mass of two objects. The only problem with it is that his equation cannot be applied accurately to an object in motion or over large distances. That was one reason why Einstein created the concept of curved space and he was correct that gravity had to be curved, but incorrect that the curvature was due to curved space.

The benefit of the theory of gravity that I described earlier is that it produces curved gravity. Proton radiation rises vertically from a planet because planets are always spherical, so on average the waves from anywhere in the sphere balance out to produce a vertical force. The density of proton waves will reduce as they spread out upwards, so that spheres of equal gravitational force exist around a planet. These spheres obviously present a curved shape to any object passing the planet.

The mass of an object equates directly to the number of protons in the object because protons and electrons are always in equal numbers and it is the number of electrons that produce the mass of an object. Thus the number and density of proton waves rising from an object match the mass of that object.

Newton's equation can be adjusted to reflect large distances between the two masses instead of a simple constant distance, and my theory can continue to use Newton's mass as the function that causes gravity.

The acceleration produced by gravity is caused by the electrons in the passing object receiving blue-shift of the proton waves, as described earlier. Thus proton radiation theory supports Newton including the acceleration of an object that is passing Earth and does not require Einstein's theory of General Relativity.

Chapter Five

THE MISUNDERSTANDING OF WAVES, PARTICLES AND REALITY

I have discussed much of this in the opening chapter. The problems seem to be,

a) If you fire electrons through two slits in a barrier they produce an interference pattern on a screen suggesting that the particles somehow behave as a wave.

b) If you fire a succession of single photons through double slits the pattern on the screen suggests that the single photons have passed through both slits at the same time, and some have reached places on the screen as if they had passed through the solid barrier.

These problems, combined with the Uncertainty Principle, which states that it is impossible to determine both the position and the momentum of a particle at the same time, led to probability theories on the most likely position of a particle, and ultimately to quantum theory, and the daft concept of a parallel universe. Instead it should have raised the question that 'perhaps we do not fully understand the relationship between a particle and its field'. I will cover much of this in the later section of Quantum Mechanics.

The key points that I have discussed earlier and perhaps the misunderstanding are,

a) A particle has no momentum of its own and must remain in the wave, which is its momentum.

b) A wave is in a straight line and this direction can only be changed if a force acts on the particle to change its direction, and the particle then changes the direction of its wave (its momentum).

Duality

The concept of duality is that light, electrons and atoms sometimes behave as a wave and sometimes behave as a particle.

Behaviour as a wave.
I think it was De Broglie who proved that if a particle has mass, then it will oscillate as it travels. But I have suggested that particles do not have mass, so my short answer is that a particle does not behave as a wave. If a wave is observed it is because its electromagnetic wave is carrying it in its own wave.

If streams of electrons are passed through double slits, one solution is that the proton radiation from the slit atoms deflect the electrons with their gravitational force. This deflection causes the electron to change the direction of its wave momentum. The effect occurs at both slits so that the deflected waves now interfere. A second solution is simply that the wave passes through both slits and interferes with itself. But the electrons must remain within the part of their wave that still exists, in effect the electrons are dragged by the wave, therefore they follow the interference and form an interference pattern on the screen. An atom will do the same simply because electrons control the movement of atoms.

Behaviour as a particle
Electrons are clearly a particle; the issue is – is light a particle?

Light is made of small packets of energy ie photons, and these are regarded as behaving like 'particles'. Perhaps this question arises because a photon has mass and momentum and this would cause its wave to behave like a golf ball.. But a particle does not have mass, so you cannot assume that even a particle behaves like a golf ball. A wave is a better description of the behaviour of a photon. Does all this really matter?

Young's Double Slits.

Now consider the phenomenon that appears to show that a photon can be in two places at the same time. Their reasoning is that the wave of a single photon appears to interfere with itself behind the slits and it can only do this if it has passed through both slits at the same time. Also, the photon hits the screen in an impossible place as if the barrier did not exist.

When a single photon passes through a slit in a barrier the wave spreads out on the far side. If this spread beam is now separated and allowed to pass through two further parallel slits in a second barrier, the waves spread again so that the two interfere with each other on the far side and an interference pattern can be seen on the screen.

The problem for scientists is that the wave beyond the slits interfere as if it has passed through both slits. And, in any event, it would not be possible to get an image on the screen in a position directly behind the barrier.

I believe the solution is that the wave of a photon must have special properties. Why is a photon able to split into two without affecting its wavelength when clearly energy has been taken away from the structure? To answer that I think one has to consider that energy can be described in two forms.
1. Level of energy. Referred to as the wavelength.
2. Amount of energy. I.e. the number of wavelength passing through an area of space every second.

A magnifying glass is an example where wavelengths from the sun are focused into a small area so that the amount of energy becomes so high it will burn things.

An orbiting electron has a specific 'allowable orbit' in which there is a specific wavelength and a fixed number of wavelengths leading to a total amount of energy in its circular field. The difference between this amount of energy and the ground state (its lowest level) is the amount that must be drawn off to produce a photon of light.

So each photon is not simply an electromagnetic wave with one wavelength, it is an *amount* of energy (several waves) all at one specific wavelength.

By that I mean the photon wave has many strands of individual waves (front to back) making up its width and the number of strands is a variable depending on the total amount of energy that is given to it at the specific wavelength of the electron.

Each strand operates in isolation and strands can be split off without affecting the wavelength of either part as it simply reduces the total energy, so in effect producing two almost identical photons.

So in my view a photon wave has a 'width' of strands that can become separated with no effect on the whole. When these bend around the atoms that form the edge of the slits in the above experiment, they are able to repel each other like ripples on a pond. The spread waves can then pass through both of the second two slits as if they were two entirely different photons, but with the same wavelength as the original photon. (Again, like ripples on a pond, except space has no water!). But all that is just a theory.

Firing a single electron through two slits produces an interference pattern exactly the same as the photon, and I suggest that the reason is the same. The electron passes through one slit, but its wave can pass through both slits and interfere on the opposite side. But the electron must remain in the part of the wave that still exists because that is it momentum, thus it must follow its wave and form an interference pattern on the screen. The electron is not really behaving as a wave, it is simply being carried along by a wave.

Quantum Mechanics.

This is not really a relevant topic for this book because, in my opinion, it is not relevant to the universe. It seems to me that scientists have created problems for themselves because of the basic errors that I have discussed earlier in the book. If you can accept the errors and my suggestions to correct them, then the need for Quantum Mechanics goes away.

QM arose because classical physics could not explain why an electron orbiting the nucleus of an atom does not radiate its energy away. But I have suggested that an electron is 'carried' by its wave, and that the wave is its momentum. So if the wave is locked into a circular allowable orbit, the electron is being carried in that circular wave, and the attraction of the nucleus is not sufficient to change the direction of the electron and its wave from its circular orbit, then there is no cause for the electron to radiate energy away. However, when spontaneous emission occurs, and the fixed orbit is broken, the electron is momentarily freed from its wave and can be pulled closer to the nucleus, until the wave is once again locked into a lower allowable orbit.

Classical physics can be linked to particle physics by Maxwell's equations – which in my view is one of the greatest contributions to physics by anyone. Using these it is possible to relate the motion of a wave of water in a pond directly to the wave on the particles that make up the atoms of the water in the pond.

The main elements of quantum mechanics are Quantum field theory, Duality, the Uncertainty Principle and Entanglement. The following is my brief opinion:-

If space is not curved and is not the source of gravity, *Quantum field theory*, vacuum energy, non-zero ground state and wave-function mathematics are unnecessary. If curved space is not the source of gravity and a charged particle does not wave or have mass, the whole basis of the wave-function seems to be invalid.

The measurement of the position and momentum of a particle that is the basis of the *Uncertainty Principle*, involves the measurement of two separate but directly related parts of an electron system. The momentum is a measurement of the wave, and the position is a measurement of the particle. If energy is used in their measurement, then, because of their direct relationship, the measurement of one must affect the value of the other, thus it is impossible to measure both at the same time. This is not a mathematical conclusion, it is a statement of measurement impossibility.

I have discussed *Duality* earlier in this chapter and suggested why an electron produces an interference pattern on a screen when passed through two slits. The electron does not behave as a wave. It is carried by a wave.

Entanglement seems to have led to long discussions between Einstein and Bohr in their attempts to explain why two photons fired together, always become polarised at 90 degrees to each other. Niels Bohr's idea, that they are able to communicate faster than light, is of course, complete nonsense, but this led to the Copenhagen Interpretation, that reality, or the actual properties of something, do not exist until you measure them, which in my view is also complete nonsense. Einstein's view, that the polarisation was decided before they were fired, is close to correct. In my opinion, what happens is that when the two photons are fired together, their proximity causes their two waves to react with each other instantly to reach stability. The electric fields will attract each other, and the magnetic fields will do the same, thus the two photons are immediately stabilised and set polarised at 90 degrees, right at the start of firing. My argument is supported by the fact that if 20 photons are all fired together, there is no entanglement. The number of waves involved makes it impossible to achieve simple pairing and stability.

But that is not a complete solution because the two photons, having moved away, change their state at the same time as if they have communicated faster than light. One theory is that the two photons become 'paired' at source so that the quantum state of one cannot be described independently of the other. Thus if one changes, the other must change.

But I think the key to the problem is that the change occurs faster than light and that means communication between the two photons is impossible, and therefore there must be a third energy influence that causes the result.

If one can accept my theory of proton radiation, then a third source of energy exists everywhere that will strike the two photons. I have suggested that these waves create gravity, light lensing and red and blue shift, and so clearly have an effect on photons.

That in my opinion is the sensible solution to the problem. The energy that causes one photon to change, will cause the second photon to change simultaneously. Nothing is moving faster than the maximum velocity, and nothing is happening that is outside the facts of physics. Only two photon states are possible, so it is a simple flip-flop situation.

But this problem seems to have led to a way of thinking that 'reality' is not decided until things stop moving, or until you look at it or measure it. A spinning coin can stop 'heads' or 'tails' so that reality is not decided until it stops. My logical brain cannot handle such nonsense! The reality of a spinning coin is simply that it will stop spinning and end up heads or tails. It is not both heads and tails at the same time. Which side is observed when it stops is a random choice. There is no requirement for any other reality.

If 'wave-function' mathematical equations are used to attempt to predict 'reality', such as to calculate the probabilities where an electron is located, (for which I have the simple answer that it can only be within its wave and the wave can only travel in a straight line, therefore we know exactly where the electron is, without any equations) then the equation introduces multiple possibilities and multiple possible realities. But common sense should tell you that these realities cannot exist.

I doubt the validity of the wave-function, but regardless of that, mathematics is a virtual device and just because an equation may provide several solutions, it does not mean that all of these solutions can actually exist. There are only three space dimensions. There are only two opposites of anything, whether polarisation or polarity.

Quantum computers

My old computer that I am using to type this book is the 0 and 1, on and off type of bits, but quantum computers can make far more and faster decisions than my old computer.

I have suggested that the waves of photons and electrons are made of many strands, and it is not that an electron or photon can be in several places, or several states, at the same time, it is that the strands of an electron field carry the electron, and the strands can be separated so that each strand can be given different properties. That means 50 strands can all make different decisions at the same time, and that is why quantum computers work so fast. An electron is reality and cannot be in several places at the same time.

None of this helps me to understand the universe. If the errors discussed earlier are corrected, is any of this required? It is a mathematical invention to solve a problem that does not exist and a branch of physics that I do not agree with and do not wish to pursue further.

Chapter Six

THE MISUNDERSTANDING OF TIME

The final misunderstanding is 'time.' We exist in a space where major events are occurring. Earth rotates in a short period. Earth orbits the sun in a much longer period. The universe is expanding slowly over a very long period. So we perceive that for these events to occur there must be time, or they would be stationary. But it works the opposite way to what we perceive. It is because these 'physics' events are occurring they cause us to think that there must be a physical property of 'time', but there is not. All the events are independent with no direct relationship. There is no continuum of time. We simply calculate how long each of these events take using a standard mathematical equation based on one day's rotation of Earth, called days or hours, or minutes. Time is just mathematics, not physics.

The fascinating concept of time is that it is said to slow down the faster you travel. If that were true it would suggest that there is no universal time. In fact I will show that there is no dimension of 'time' at all, it is simply a mathematical invention and it has no presence in physics. There is just 'existence' or 'now', and movement we see in front of us. We compare that movement of an object with the movement of a clock and conclude that time has changed. But it is motion that slows, not time.

Time is not a dimension because,
- There is no evidence that it is a dimension.
- There is no 'actual' time, there is only the time on a clock, or the relative time caused by the delay in the image of a clock reaching a distant observer.
- The dimensions of space are clear because they allow us to move easily up, down and sideways. We cannot go backwards and forwards in time. If we cannot go there it is not a dimension.
- It is not required for the universe to function. It exists only as a perception in our brain as a result of things moving in front of us.

If time were a dimension we could go back to Friday and fill in

Saturday's known lottery numbers! And unfortunately it takes some of excitement out of sci-fi movies.

The story of two twins where one travels through space and the other remains on the ground for 10 years has the wrong answer. Both twins are 10 Earth years older, but the space traveller moved more slowly, his heart beat very slowly and his watch moved slowly. He probably aged more slowly if he did not die. Time on the spaceship did not really slow.

Time is just man's invention to record how fast things change. Things move in the universe because of energy and it is this movement that causes us to invent the concept of 'time'. But 'change', including the changing position of the hands of a clock, does slow the faster the object or clock moves.

We all can see the sun rise in the morning, move across the sky and set over the horizon at night. We all see the hands of a clock move around hour by hour. So we can all see that time passes. In reality, time is just a perception.

We measure the time night changes into day. The time it takes from being at the start, to being at the finish of the 100 metres sprint, but the second hand of the clock is also changing from zero to about ten seconds, so time is just man's invention to measure change and plan future actions.

You will say "surely time exists because I need it to get to the station to catch a train", or "My birthday was a month ago, so time has passed". NO. We *choose* to measure birthdays in days, and we know we can only change our location from home to the station by using a few minutes shown on the hands of a clock, which is the same device that the train uses to arrive at the station. Man invented the clock, but the universe has no need for it. We use time in physics to calculate the rate of change, or velocity, only because man has a desire to do so. We wish to know the speed of light.

Motion Dilation

I suggested in chapter one that time dilation, whilst it can be argued as being correct, a more precise term would be motion dilation, or retarded motion.

The reason is that as velocity increases it becomes harder and harder for electrons to convert energy into the larger wave, or momentum, that they must have to be able to accelerate, and electrons control the movement of atoms, and so that the 'inner workings', or the mechanism of all objects, will slow down the faster the object is moved.

So it is the clock in a spaceship that goes slow, not time itself.

The only factor is that motion occurs slower the faster you travel.

Because no changes can occur at the speed of light, we perceive that time has stopped at the speed of light. (The electrons cannot move / change, and they control atoms). At lower velocities the process of creating wave can be accomplished very quickly so a particle can change status quickly. We interpret this inversely that time has moved faster.

If an electron cannot move forward then the mechanism of a clock of whatever type, cannot work. When the clock slows we say 'time slows', but what we are really observing is that the rate of change of everything, including the clock, slows, and that really means the ability of electrons to change slows.

The delay in creating a wave might be explained by the fact that a phone call from London to New York has a small delay and this is because the wave on each electron in the circuit transfers to the electron in front to achieve equilibrium of energy, but it takes time for the front electron to receive and then convert that energy into a wave of its own.

One further point is that the 'energy-mass equivalence' equation $E = mc^2$ produced by Einstein can also be used to define the process in which energy is converted by a charged particle into mass and velocity.

It is clearly an electrical conversion process and all processes must take a nanosecond of time. The conversion also requires a charged particle to perform the conversion, so the equation really says that if energy is applied to a charged particle, and you wait a nanosecond, the particle will accelerate and produce a wave. But it is the nanosecond delay that is the secret of the rate of change, and therefore the concept of time. So, in terms of the process, perhaps Einstein's equivalence equation could be written as,

$$E\,t = mc^2$$

At the speed of light there is insufficient time for the rate of producing a field to produce a longer length of material so no increase in velocity or movement is possible and this means that no 'change in status' can occur in the direction of motion.

If you think deeply enough into this you can see that people walking (obviously at very slow velocity) must actually move in extremely tiny jerks because movement is only possible when the energy of the muscles causes an electron to produce higher momentum!

Gravity does not cause time dilation.

At this point, I need to explain the Shapiro effect of gravitational time dilation, or how gravity appears (wrongly) to affect time, and why GPS systems must make a time adjustment of 38 microseconds to bring their clocks into line with those on the ground.

GPS uses radio waves. These are negative, the same as light waves. Experiments have shown that light waves take longer to pass through a strong gravitational field than in free space. But the speed of light is always constant, so one may think that time must slow in a gravitational field, but that is incorrect. I explained this earlier that the real reason is that gravity waves are positive and so cause light and radio waves to have to manoeuvre around them, so making their journey longer. Time does not slow down or speed up. The effect will be exaggerated by the density of the proton radiation around a massive object where their gravity is strongest.

Our perception of time.

The concept of travelling back in time is nonsense as all previous events no longer exists.

If time is just lots of very fast and very tiny changes to electrons, why do we think it goes on forever? What is 'forever'? Why do we think there is a past and therefore there must be a future?

The answer is threefold,

a) Because we have a device called a memory.
b) Because we can look back in time to distant stars..
c) We see isolated events as being somehow related because they happen so fast.

The terms 'past' and 'future' are just words that man has invented in order to communicate. There are no such places or time zones in the universe, there is only 'now'. Our memory is able to store all the changes that have occurred and we can visualise what it was like when that energy flew past. We also have photographs to remind us. So we assume that time must also be stored somewhere in the universe, but it is not. It is only stored in our memory. It is gone. The past no longer exists. We cannot travel to it. The changes that have occurred cannot be reversed.

Because we can recollect the past we think perhaps the future is already mapped out and we will simply move into it. We are safe to assume that there will be a future because the world has not come to an end, but in physics terms, there is only 'now', our current existence. We can make tomorrow's 'now' anything we want it to be by preparing for it in today's 'now'! Nothing is mapped out ahead of us and we can decide our future for ourselves.

Time exists only as individual particles of energy at the smallest atomic level, and what we interpret as a single thing called time is in fact billions of isolated electrical processes that create their own time to cause minute changes. The time they create is simply the process of turning energy into velocity. The momentum is the store of energy that allows an object to continue to move, but time is not required for the movement. Man invented the concept of time just so he could measure velocity.

We do not live at the atomic level, we are at the planet level: we do not see these tiny processes; we can only observe the result. We perceive a continual 'single universal time' because we are observers at the macro level.

Some say that if we could travel faster than light we could go back in time, but there is no logic in this. Even if we could travel instantly to another location we would only get to 'the present time' for that location.

The image of distant time

This does not really require explanation. Every atom produces snapshots of light extremely quickly. Their source is the electrons orbiting the nucleus of every atom which produce a specific colour, and the emission that leaves the atom is a photon. When the individual snapshots from all the atoms around us are added together, a complete coloured image of something is produced, exactly like the pixels of a camera or television. These images travel at the speed of light, because they are light.

So by the time the image of a star reaches us it is already several years old. And once that snapshot of image has passed us we can never see it again, nor can we step into a spaceship and try to catch up with it because we cannot travel faster than light. So it is impossible to see images of stars that passed by centuries ago. We cannot travel back in time.

The time cone of the universe is a good example of looking back in time, but which does not show that time is a dimension, it is just an image.

But I just used the phrase 'several years old'! What does that really mean? Such a measure of time is simply man's invention because the image that reaches us is no older or different from when it left its source. We have used the period of rotation of Earth around the Sun as a time comparison to show how long the image has taken to reach us. But the image itself did not experience any time and that is what matters! At the speed of light, change (that we interpret as time) has stopped.

Time will eventually stop.

The question 'when did time start' becomes rather meaningless if the universe has no universal time. However, it is fair to ask "When did the first thing start to release energy to give objects velocity", and that is explored further in a later chapter on the creation of the universe.

The next obvious question is 'will time go on forever'? As there is no 'time' in the universe we should rephrase that and say "Will change go on forever"?

My answer is 'No'.

The energy on electrons that allows 'change' comes from the stars and these will eventually run out of energy, but the real problem is proton radiation. This requires energy stored within the nucleus of atoms. This energy will eventually run out as protons decay and become exhausted, so radiation will stop. The space dimension will start to disappear.

Electrons will become unable to change their energy. Change will stop. That means our interpretation of time will stop. Time had a beginning and will have an end simply because the universe had a beginning with the Big Bang, and will have an end when proton energy runs out.

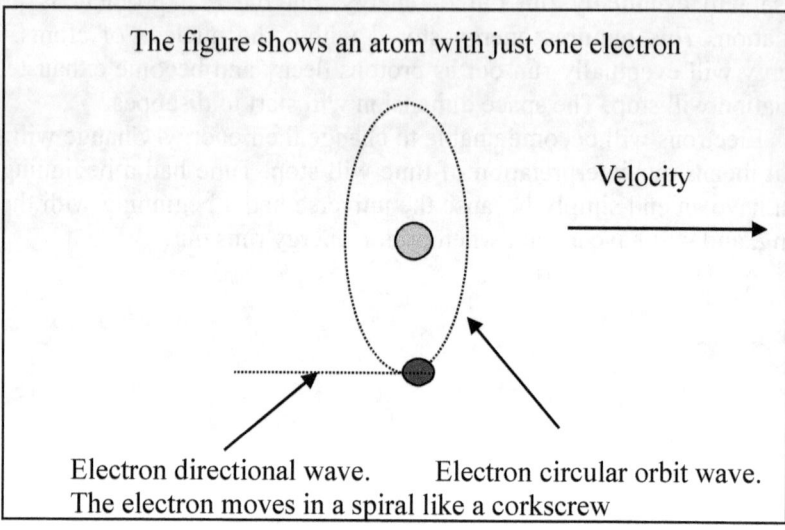

- The position and movement of an atom is controlled by the electrons. The nucleus will be dragged where-ever the electrons move.
- An electron must have a wave in the direction of motion for it to be able to move, and for the entire atom to move.
- Creating the wave from energy is a process that takes time. It is $E = mc^2$
- The rate of converting energy into wave is constant, therefore the greater the amount of energy to be converted, and the higher the velocity of the atom (and its electrons) the longer the electrons take to produce the amount of wave required, so the longer it takes the atom to change in the direction of travel.
- At the speed of light it cannot produce a wave fast enough and so change stops in the direction of travel.
- We perceive 'change' as time, so we say 'time stops'. Because electrons cannot change, atoms cannot change, so the hands of a clock cannot move, so the clock stops.

The reason clocks slow at high velocity.
(This is not due to Einstein's Special Relativity theory.)

Chapter Seven

THE FABRIC AND SIZE OF SPACE.

Having straightened out some of the misunderstandings and the existence of proton radiation that fills every cubic centimetre of space we can now go forward and explain the universe itself.

A volume of space cannot exist unless there is material that makes up that volume dimension. It cannot be just an empty geometric volume. So at the time of the Big Bang, there was no 'material', and therefore there was no space. There was no universe; it was 'nothing'. It did not exist.

Similarly we can say that if something produced the material that formed the dimension that we call space, then something is still producing that material because the universe is growing larger.

Consider a hypothetical case that I thought through in the very first few days of this project in 2003 to understand the universe and how atoms work.

We believe that the boundary of the universe is the furthest point at which light from the galaxies and the Big Bang have reached. Now suppose there was a spaceship that could travel much faster than light. What would happen when it reached the boundary of the universe? My conclusion is that, because the spaceship produced light and proton radiation, it would extend the boundary of the universe along the path it took, so in effect it would produce its own little universe.

But now what happens after travelling a few hundred miles further. There is no light from the galaxies or the Big Bang because that light has not reached where the spaceship now is, so it is in total darkness and the universe cannot be seen. Firstly, that means the captain has no idea where the universe is and does not know in which direction to travel to return, but it also means that all the electrons in the atoms of the spaceship cannot absorb light and must fall to their ground state. So spontaneous emission is now impossible and the captain will not even see his own spaceship.

So does that mean the spaceship loses its little universe and disintegrates? My conclusion was no. Atoms cannot disintegrate or disappear just because they are not receiving light waves. Clearly the spaceship atoms produce their own waves and so produce the little universe dimension that the spaceship requires.

And, if the spaceship stops travelling, the captain's watch would show that relative motion is still possible even though the spaceship is outside the universe! So that means that the little universe of space is all produced by the atoms of the spaceship, or in other words, every atom produces its own universe and it all comes from the nucleus because the electrons are inactive.

But of course that is all hypothetical nonsense and does not prove anything, it only shows my way of thinking 'outside the box'. However, if we assume that the universe did not exist prior to the Big Bang, then the volume that we call 'space' did not exist. It was 'nothing': no volume, no dimension. Clearly something must have produced the volume and that is probably the Higgs Field, but does it still exist? If not something must still be producing the volume, and if that material is still being created, it must come from the most abundant item in the universe – atoms. Because the proton must decay, it is reasonable to think that the material filling the universe is proton radiation. My logic above suggests that such waves must exist, and must be positive for the rest of the universe to work as it does.

We know that atoms containing 100 protons radiate positive Alpha particles, so it is reasonable to assume that all protons radiate energy, but Alpha rays are not simply electromagnetic waves, so they are not really proof. What is more important is Grand Unification Theory that predicts that protons radiate waves.

But if a wave produced by a proton is 'space', then one can conclude that the waves from every charged particle are a piece of space, including a light wave. And if an electron produces a wave that is space, then that wave exists so that the electron can move. I.e. the space is unique to the particle that created it, and is of no benefit to any other particle.

The key point in my theory is that the proton radiation is radiated and travels in every direction and forms the space dimension, but the wave is invisible. The reason we can see objects in different positions in space is because of light waves, and these too help to form the dimension that is space.

But neither light waves nor proton radiation enable an electron or an atom to move because every particle must make its own space (its wave), but stars, planets and entire galaxies do exist in their own particular location of proton radiation, and one can say that it is the proton radiation that provides the 'relativity' and the 'field' that enables electrons and atoms to move. If there is no frame of reference, velocity is impossible.

But why does a particle produce a wave? Presumably it is just what energy with velocity does and we are fortunate that this creates space. My conclusion is that a particle cannot exist at all unless it creates a space for itself and that is why protons must decay and electrons must have velocity.

I will discuss momentum later, but if we accept that a stationary charged particle has no momentum or mass, and that when energy is added, the particle stores this energy in the form of an electromagnetic wave and it is this momentum in the wave that keeps the particle moving in space. Thus a particle can never become separated from its wave or it would stop moving. So it follows that the wave is an essential requirement for a particle to be able to move. It is now just a short step to say that the wave, in addition to momentum, is the dimension of space that a particle must have in order to move.

Thus the dimension of the universe, or 'space', is the sum of all the proton radiation, light waves and any other waves passing through it. It is interesting that a wave can travel through space as if there was nothing there, yet to a particle, space is like solid rock. It requires a wave to drill a hole through it! So it seems to show that a wave is a dimension necessary for travel.

Space only becomes three dimensional because billions of these waves of radiation are moving at the speed of light from every atom and in all directions, and one group of waves is immediately followed by another, so that space appears to be a continual volume in three dimensions.

They do not collide because their charges will repel each other, but there cannot be gaps between these waves as the gaps would be 'nothing', and nothing cannot have a dimension. The gaps would close immediately.

If the universe of space started as 'nothing', then proton radiation is capable of travelling through 'nothing', and this helps to understand that there is no volume, or space outside these electromagnetic waves, and so outside the boundary of our universe there is nothing.

The radiation from every galaxy will travel in a straight line, at the speed of light, forever. Most will hit other stars, where the energy they carry will be released as a gravitational force on the electrons in their atoms. Many will not hit anything and will just keep going. At a far distance from the galaxies, where the density of the waves become low, relativity between the electromagnetic field of two waves will become lost. The waves are no longer touching, so the dimension of space they produce breaks and is lost. The waves stop moving. This is the boundary of space.

One would think that perhaps each individual proton wave would continue by itself at the speed of light as a tiny blip of energy, but without relativity, velocity is meaningless because there is no reference point to detect movement. The wave cannot move. It is in its own little universe in 'nowhere' or 'nothing'. I will explain in the chapter on Conserved Relativity what I believe actually happens.

If our universe is surrounded by 'nothing', which means that there is no dimension because there is no proton radiation, then there cannot be another universe because the zero dimension would cause the two universes to combine into a single universe. And if there is no 'time', as I explained in an earlier chapter, then there cannot be a second universe occupying the same space as ours, but in a different time. So the 'Multiverse' theory cannot be correct, and I see no reason why it should even be considered.

However there is one further concept that is mildly interesting. Instead of assuming there was only ever one Big Bang, what if the universe contracts back to a dot as I propose in a later chapter, and what if in 'nothing' outside our universe, charged particles were being magically created every few minutes all the time. These particles would immediately come to rest at the boundary of our universe. Then when our universe collapses back to a dot, all these new particles will become part of our universe. Thus it is possible that, over zillions of years, our universe started as just a single proton or boson, and has grown by the addition of trillions of other particles or protons being created in 'nothing' and absorbed into our universe at every contraction, so producing our universe at its current size? That is just a concept that I will not pursue further!

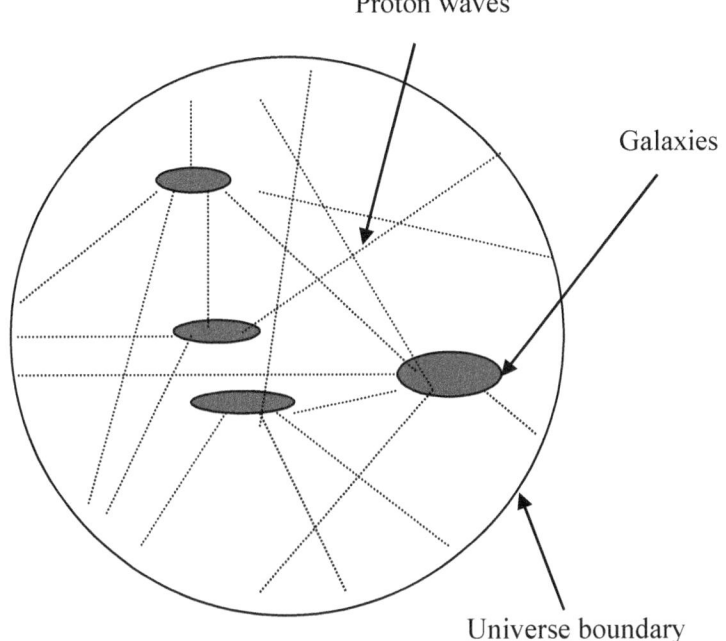

THE FABRIC OF SPACE

Millions of proton waves leave all stars and planets. They criss-cross through the universe and cram every cubic centimetre full, so creating the volume or dimension that we call 'space'. Outside the universe is 'nothing'. No waves, no dimension. It is hard to imagine but the location does not exist! (I am unable to draw 'nothing'!)

Space must eventually disappear.

It seems clear that when all the stars have burned out their energy and all protons in the nucleus of atoms have fully decayed, radiation will not be produced and space will instantly disappear. If there is no radiation there cannot be a dimension that holds stars and planets apart, and the universe returns to nothing. (Note that when I refer to 'proton energy' I believe it is actually the entire energy of the nucleus that expires but does so via the positive proton).

If, as I discuss later, there is a law that states 'relativity must never be lost', then the radiation will never have left the universe. Everything, including the radiation and photons from billions of years previously, will instantly contract to a single point when the space dimension disappears, and the Big Bang will be repeated. Particles with the same polarity will again be pushed apart. This logic will be expanded in later chapters on the contraction, the Big Bang and the creation of the universe.

Finally, just a word on some absurd theories that have evolved. Scientists have a theory that there may be ten or eleven dimensions rather than just the three that we all understand and this theory came about as a result of mathematical solutions to problems. The theory evolved because when quantum theory was applied to electromagnetic waves, infinite values for the mass and charge of the electron were produced. As I understand it, this problem led to the use of Grassmann maths and symmetry of particles. Then, using this form of mathematics, several dimensions became possible in order to make the problem equations balance. But mathematics is a virtual subject and one cannot accept all the answers as being reasonable in the real world. And you have to be certain that all the assumptions used to create the equation are correct or the equation is meaningless, and that is what I believe has happened.

The Mysterious behaviour of particles in space

Proton radiation comes from the sun, moon, stars, trees, and clouds, in fact there are trillions of waves travelling in all directions in just one cubic centimetre. Any negative particle travelling through this matrix of positive waves will be buffeted by the waves in all directions a million times a second, thus such a particle will wobble whilst the momentum of its wave keeps it on a roughly straight line. Such transfers of energy, which I discuss later, may also cause an electron to accelerate or slow down.

In my opinion it is this matrix of forces and transfers of energy between the positive waves and the negative electrons that could cause a random particle in space to disappear and re-appear. Thus particles are quite possibly obeying the normal laws of classical physics and so it seems likely that a particle does not possess some unusual powers that only quantum physics can define?

Chapter Eight

DARK MATTER, DARK ENERGY AND HOW GALAXIES ARE HELD TOGETHER

Dark Matter

I have never really understood what is meant by the term 'matter'. If it means protons or atoms, then clearly there are not any missing or they would have been seen. If it means a source of mass or gravity, then the answer is easy. If it means a force that prevents a star from flying out of the galaxy, then that is not due to matter, I believe it is due to gyroscopic precession.

In my opinion no additional matter is required to prevent stars from flying out from the galaxy, it is a matter of complex gravitational effects. But first I will just deal with the fact that scientists have identified that there is some missing mass, or mass that scientists cannot identify in the universe.

This has been discussed earlier and in my opinion it is just proton radiation waves that completely fill every cubic inch of space to provide the dimensions of space. The waves have mass because they are a field with velocity, and both are a source of gravity, wherever they are in space, so they are similar to 'matter', but 'matter' does not produce gravity, it is primarily the proton radiation from matter that produces gravity. If you vaporised matter with a nuclear bomb, the gravity they produced before you blew it up would continue to exist as it travelled further out in space.

Dark Energy.

The term 'dark energy' arose because, Hubble discovered that whilst gravity was a force pulling all matter together, there seemed to be another unidentified force in the universe that was pushing galaxies outward and causing the expansion of the universe. A kind of 'anti-gravity' that made up 75% of the universe. The galaxies were found to be accelerating outwards, and a further phenomenon arose that everything in the universe is moving away from everything else.

My conclusion is that both of these are caused by the same thing. Proton radiation. The material of space behaves like rubber! It is compressed around large objects where it is created, but forces itself to expand as it moves away from the object.

Trillions of waves will be created by the protons in a star or planet and each element of the positive proton radiation repels its adjacent elements and pushes them away as soon as it emerges from the surface of the star. So as the waves move away from their source at the speed of light, the waves separate. That means the dimension of space broadens and expands and it carries with it all the stars and planets in that expanding volume of space.

Now what we observe is simply galaxies and stars moving away from us, but what is actually happening is that the location of the universe in which the galaxies exist is being moved by the expanding radiation that make up space.

So the dark unidentified energy that is causing galaxies to accelerate away is the repulsion of each of the elements of proton radiation as the try to expand like a compressed rubber ball.

These waves are being created and radiated all the time, so space is being created all the time. The expansion will continue and everything will become further apart.

But the universe has a boundary and that is where the dimension stops. It is the furthest point where proton radiation has so far reached and without the waves there is no 'material' and no dimension. Beyond that point there is nothing. No dimension. The location does not exist.

'Nothing' has no volume or relativity. So the size and shape of the universe is the size and shape of all the galaxies within it, plus the distance a number of streams of waves can go before their electromagnetic waves lose relativity with each other.

But this does not explain why stars in galaxies travel faster than they should, or why galaxies do not fly apart. So now I will suggest an explanation for that, firstly by explaining how gravity causes precession, then how gravity causes stars to rotate, and lastly how I believe all this prevents galaxies flying apart.

If my solution is correct, I do not think that there is any dark energy other than proton radiation described above. And if 'dark energy' or 'Dark matter' are simply terms to equate to 'missing mass', then I have already explained that. It is photons and proton radiation.

So the entire universe contains mass, but this does not explain what prevents stars from flying out of their galaxy when gravity by itself is not strong enough to hold them in. So I will move onto 'precession' and how I believe it holds galaxies together.

Gyroscopic precession caused by gravity.

This section may seem an odd inclusion but it is necessary to understand why a gyroscope does peculiar things in order to understand Dark Matter, and by Dark Matter I don't mean the 'missing mass', I mean the force that prevents stars flying away from a galaxy. The force involved in galaxies is, in my view, simply gyroscopic precession.

For those who do not understand gyroscopes, we all know that a toy gyroscope does peculiar things. When spinning on a horizontal axis, with freedom of movement up, down and sideways, it can support itself at just one end of the shaft while it rotates, or precesses with its shaft around that support. And if we block its path so that it cannot precess, then it will rise vertically as if by magic. These peculiarities are caused by gravity. A gyroscope will perform its stabilising effect perfectly normally in space as that action is simply angular momentum.

We also all know that a 'spinning top' toy will stand vertically until friction causes the spin to reduce to a level where the 'top' starts to tilt. At that moment precession begins because gravity starts to have an unequal effect on the electrons in the atoms. In outer space the top would be unable to tilt as there is no 'vertical', and there would be no precession.

To clarify these peculiarities, spinning a heavy disc or toy gyroscope on a horizontal shaft in a gravitational field, with freedom to move, produces two effects,

- The disc is able to support its entire weight at the opposite end of its axle. This is possible because additional mass is created in the disc in the horizontal plane through the centre of the disc at 90 degrees to the axis, so that it is the angular momentum from this 'stationary' concentration of mass that forces the disc to raise its shaft into a horizontal position.
- The disc rotates, or 'precesses', around its support, horizontally.

These two effects are directly related. There cannot be one without the other, and because the source of both is gravity, maximum precession can only take place along a horizontal plane, and when the rotating disc is vertical and spinning on a horizontal axle. Reduced precession will occur at angles between vertical and horizontal where gravity has a reduced effect.

Consider a heavy disc or gyroscope mounted horizontally on a shaft as shown in the drawing. The disc is made to rotate rapidly by a motor.

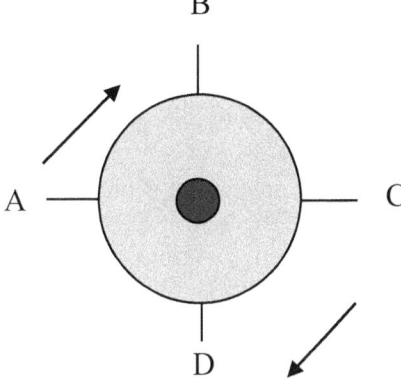

A heavy disc rotating clockwise about a horizontal shaft

It is known that if downward pressure is applied to the shaft the entire assembly will precess, but this is not due to the shaft, it is because all of the force passes to the electrons in the disc. Gravity has the same effect and this behaviour is further proof that gravity only effects electrons.

The disc is of course made of atoms, and atoms are made up of electrons orbiting the nucleus. Gravity is a positive wave and therefore is only able to have an effect on negative particles. The only negative particles in an atom are the electrons, thus gravity only affects the electrons and has no effect on any other part of an atom.

Proton radiation will of course add momentum to all the electrons in the disc. In segment D-A-B, gravity momentum acts in the opposite direction to electron motion so reducing electron velocity and momentum.

The converse is true for the other half of the disc where gravity adds momentum in the same direction as electron velocity, so increasing velocity and momentum.

This unbalanced momentum places the moment of rotation at a position to the right of the axle, so the axle precesses horizontally to the right at a speed that puts the new axle at the new axis of rotation.

The velocity of precession places new momentum at points B and D, however the momentum at B is higher than at D and this changes the moment of rotation of the disc, so causing the axle to move upwards until it is horizontal where the centrifugal force from precession is at its greatest.

Clearly both of these are interdependent. Without precession the axle cannot be horizontal, and without gravity, momentum and inertia cannot cause the disc to precess.

We can also see that if precession is physically prevented the disc is forced to remove the source of the 'out of balance', i.e. gravity. It does this by rising vertically until the disc is rotating horizontally on a vertical shaft. Or in more technical terms, the excess momentum drives the disc upwards until the influence of gravity becomes zero.

In short, the velocity of precession is set by the speed of spin. Once this equilibrium of momentum has been restored by precession, the disk will perform exactly like a stabilising gyroscope, as if there were no gravity and precession. This means that spin must always occur at right angles to the axis. This may seem obvious but it is vital in understanding how galaxies work, as now follows.

Now I will return to galaxies and gyroscopes.

Galaxies are the most complex mechanism in the universe and it took me several hours over three days to unscramble it because there are three separate gravitational forces at work.

We know that gravity from the centre of a galaxy, usually a Black Hole, pulls stars inwards. This force performs the same function as the axle of a gyroscope and so every star in a galaxy has its own 'virtual axle'.

We also know that the universe itself has central gravity due to all the galaxies within it. This gravity performs the same function as Earth's gravity in the example above of a gyroscope.

The third force of gravity is just that of neighbouring stars which will tend to pull other stars on the same orbit towards the centre of the galaxy.

I discuss in the argument below why all stars are spinning. The planets around a star, like Earth orbiting around the sun once per year, are orbiting in the same direction as the spin of the star. If that were not so, how did the planets in our solar system fly out from the sun?

Finally, now assume that the flat spiral shape of every galaxy is orientated at 90 degrees to the gravitational pull from the centre of the universe. (Or it may be to a nearby galaxy that may have an even stronger gravitational influence) Now every star in a galaxy behaves as a gyroscope!

We can now see that it is not normal angular momentum that is at work in the rotation of a galaxy, it is assisted by gyroscopic precession. The 'gravity virtual axle' of each star is caused to move and precess around in a circle, and each star has pulled its neighbour into place over millions of years, to achieve a massive state of equilibrium in a giant spiral.

At this point things get subtle and complicated and you will doubtless argue, but I believe I am correct!

You will say that a star of mass m, and velocity v, has a momentum of mv regardless of whether it is spinning, and if that momentum produces a centrifugal force greater than the gravity within the galaxy, then the star should fly off, unless gravity is increased by Dark Matter.

The rest of this chapter is probably the hardest part of the book to understand so I will take it in small pieces. First the reason all stars in a galaxy are spinning, then the formation of a galaxy and finally the reason I believe the galaxy is held together without any extra Dark Matter.

The reason stars in a galaxy must spin.

Firstly, the following mechanism does not apply to every planet or moon because their atoms have too many electrons all orbiting in different direction so their masses prevent the limited momentum having an effect. But Mercury and Earth are so close to the sun and gravity is so strong, that the mechanism does apply. It is the reason Earth forms an equator.

Returning to galaxies, over a long period of time, the few electrons in the hydrogen atoms of a star will adjust their orbit so that the star is perpendicular to the motion around the galaxy. That means that their orbit will be aligned towards and away from the centre of the galaxy, and that means that it is directly aligned with the central gravity of the galaxy. So here is my theory.

- The answer for spin is gravity and it is the same mechanism as the gyroscope.
- The field or wave of the orbiting electron in an atom of a star is its circular orbit itself and it contains all the controlling momentum that is in an atom because the nucleus has no role.
- Neighbouring stars in the galaxy will pull these electrons so that they are in the same plane as the spiralling galaxy
- This means that the electrons orbit in all the atoms in the star in a direction exactly in line with the centre of the galaxy, and that means in line with the central gravity pull of the galaxy.
- Only electrons that orbit in this direction will speed up one way and slow down the other way (towards and away from the galaxy centre)
- Now consider that each of these electron orbits will behave exactly like the entire disk in the drawing. Gravity will either pull and increase electron velocity and momentum, or pull and decrease it depending on the electrons position in its orbit.
- These changes in velocity of electrons within atoms that have a fixed overall mass means momentum is changing as velocity changes and remember electron velocity is not constant around each orbit.
- It is this unbalanced momentum that causes a shift in the centre of orbit of each electron so causing the axis of each electron, and the nucleus it orbits, to move in a plane in line with the miss-balanced orbits.
- The axis of rotation of each electron orbit is approximately a 'virtual axis' into the centre (of gravity) of the star. Thus the electrons must move to a new virtual axis to rebalance momentum and so push all their atoms one way. (Which way depends which way all the electrons are orbiting, and they should all be the same in every atom and in every star).
- As all atoms are joined together, the star starts to spin due to precession in a direction that is at right angles to the gravity I.e. around the galaxy.

- So, although we might describe a star as 'spinning', it is actually 'precession' about its central axis due to the spinning of electrons in its atoms.

The formation and rotation of a galaxy.

- Suppose that there are now hundreds of stars (or clumps of hydrogen) that gradually move closer together by their gravity to begin to form a galaxy, and that a strong gravitational source forms at the centre. They will all begin to spin in the same direction (at right angles to the central source of gravity) for the reasons discussed above. The situation is now exactly as shown in the drawing of how a gyroscope works,
- As each star gets closer the entire star must behave as a gyroscope, but, the gravity of the centre of the galaxy now acts as the 'virtual axis' of spin.
- Precession must now occur due to the spin of the star, and the virtual axis must move to the right at a velocity that rebalances momentum.
- So the star begins to orbit around the galaxy centre and gradually all the other stars do the same.
- All the stars will interact their gravities until they form a structure that is in equilibrium, and that is the spiral galaxy.

Of course as the stars rotate around the central star their electrons must radiate energy and so their momentum will slowly reduce. The spin has its own momentum and as this is the source of precession it will tend to restore that lost by rotation, but slow its rate of spin.

Now here is the subtle part about how galaxies are held together,

- When the rotating (precessing) star is firmly into an orbit around the central star, it will behave exactly like a stabilising gyroscope.
- That means that the spin of the star must remain exactly at right angles to the 'virtual axis' that is gravity from the central star. Nothing will shift such a massive spinning object from that requirement.

- The same gravity of this 'virtual axis' will also assist this because it will attempt to cause the leading edge of the spinning star to turn inwards as the star processes. (It is the same mechanism that causes stars to spin).
- So the star is unable to move outward to a wider orbit, not just because gravity holds it in orbit at its current angular momentum, but also because - if spin must remain exactly at right angles to central gravity – the path of exit would have to be a curve, not a straight line of momentum. That means there would need to be substantial acceleration, not simply an existing high velocity. But that is impossible! To achieve acceleration would require that the spin of the star would have to increase in order to cause precession to increase, or alternatively extra energy would be required from somewhere to cause acceleration. A faster spin is impossible because the massive momentum of spin is controlled by the gravity of the universe, as discussed in 'the reason stars spin'.
- The converse applies to stop the star moving inwards.
- So the obvious conclusion is that no matter what the velocity of rotation of the star around the galaxy, its spin prevents it changing orbit.
- Neighbour stars in the same orbit will also tend to pull a star inwards.
- That means there is no requirement for Dark Matter to hold galaxies together.

There is still an unanswered question: why do some stars orbit in the galaxy at a faster velocity than we would expect, as it is this that leads scientists to think there is an extra source of gravity?

I believe the answer to be somewhere in the above. For example, either some stars force other stars into the wrong orbit, or some stars have less mass than others, or it may simply be caused by the precession itself.

The whole mechanism of the galaxy will be slowing down as stars lose their spin, and one by one they will be sucked into a Black Hole at the centre and this will raise gravity regardless of Dark Matter. This is probably why some Black Holes exist.

Chapter Nine

WHAT IS ENERGY?

What exactly is energy? We know that it is what we need when running to catch a bus. We know it is released in an explosion. We probably know about 'kinetic energy', due to velocity, and 'potential energy', when something is not in equilibrium, but what is it? We cannot see it, or touch it, but we can be hurt by it if it causes a bang on the head.

At first you may think energy comes in several different forms. E.g. heat, light, a push, a pull, height, voltage, gas, petrol etc. In fact all energy is electric, and the entire universe is electric and the energy comes in the form of potential energy (a difference in charge voltage between two particles) and kinetic energy (a velocity of a particle with mass). Everything in the universe is made of various levels of charge and electromagnetic field from positive or negative particles of electricity.

Einstein showed that energy $E = mc^2$

I have slightly modified that to include light as $E = mc^2f$

(Where f is frequency)

But these equations are just the result caused by some other source, so I will first discuss electromagnetic waves and charged particles.

Waves and Particles

There is no website that provides a description of the structure of a charged particle, so I will give you my logical approach to understanding waves, and then particles, and I have added this because such an understanding is needed to be able to understand why a positive field can cause an electron to move backwards, as suggested in chapter four on gravity.

Note that I may be totally wrong. I offer no proof or mathematics, and my opinion is based only on logic and physics.

The Structure of Electromagnetic Waves

$$\text{Energy, } E = mc^2 f$$

f is frequency as discussed in chapter three. Mass is the length of field, also discussed in chapter three. So energy is simply a wave travelling at the maximum velocity, possibly with a frequency greater than one.

It may be that a field is either positive or negative depending on its source, but there is no such thing as an isolated magnetic pole, so a structure such as a wave with a North Pole perhaps must always have a South Pole.

Maxwell's equations make it clear that a magnetic wave and an electric wave interact to produce momentum, so I don't think it can be possible to have an isolated positive electric field and wave; there has to be a negative within the electric field and wave.

Therefore one cannot say that a positive boson has a positive electric field or wave, one can only say that the lead edge is positive, and it is that polarity that Maxwell's equations use to drive the wave forward with momentum.

Thus I think a wave can be either North-South or South-North, and either positive-negative or negative-positive. But certainly it is the difference in lead edge that is suggested to be the reason gravity causes an electron to move the opposite way. The momentum is reversed as I explain further below.

The Structure of Charged Particles

If we now look at charged particles. These are energy, and energy is a wave travelling at the maximum velocity as discussed above, therefore a charged particle is probably a tiny wave with the two ends joined to form a circle, and rotating at the maximum velocity.

My definition 'rotating' is the classical term for the spin of an object. My use of the term 'spin' is the quantum terminology of spin. I suggest that the spin that is observed is simply a reflection of the frequency within the rotating wave, which must be an integer so that there are a limited number of allowable particles.

The spin, or the frequency within the rotating wave, is fixed and cannot vary unless substantial energy is absorbed.

I think the title of 'charged particle' is not helpful as, in my opinion, it does not contain a store of energy that can discharge with a spark and still exist. The only store of energy is in frequency and that cannot be discharged. If a particle discharges by touching its opposite particle the two will disappear and become something else. So a particle is not 'charged up', it simply displays a polarity, and the word 'displays' is important. 'Electric particle' may be a better title.

Now if it is true that the circular wave must contain both a North and a South, a positive and a negative, then all particles are the same structure. The only difference between a positive and a negative particle can be in their direction of rotation.

The direction of rotation will determine the lead edge, and the lead edge, driven by Maxwell's equations, produces the polarity. Thus an electron is the same structure as a positron but it will rotate in the opposite direction, so producing an opposite lead edge and giving the appearance or display that it has a different charge.

But whilst the tiny circular wave is mass and momentum, it is for internal use only because the circular wave is a fixed structure and does not form external momentum for the particle, nor is it rest mass or the cause of resistance to motion. Any external wave touching an electron cannot become absorbed into the tiny circular wave as that would try to change the frequency, and that is not possible. It must remain external simply as available energy.

The relationship between the two energies
The cause of gravity

So it is the lead edge of the structural circular wave of an electron that determines how it will react to the incoming energy of a positive wave.

Any additional positive energy added to an electron cannot form part of the particle's negative structure. It must remain as a separate wave. Thus it must become the momentum of the entire particle. So the question is, what will the particle (electron) do with this arriving positive field? I suggest that the rule is as follows,

If a particle is rotating with a wave that has a negative as its leading edge then the wave of the entire particle (its linear momentum and direction of motion) must also have a negative as its leading edge.

Thus an incoming wave that has a negative leading edge will be absorbed by the particle's wave in that polarity sequence and the particle will move in the same direction as the incoming wave.

If the incoming wave has a positive as its leading edge, the particle will absorb the wave in its polarity sequence, but when that sequence is completed the particle must use the negative as its leading edge and to do that requires that it travels in the opposite direction to the incoming wave.

So that is why a positive field can 'pull' a negative electron and create gravity. Remember that the internal wave has no resistance to motion.

It may help to give an example; Let us say that the lead edge of the circular wave within an electron is negative electrically. And say that the positive wave that touches the electron has a positive lead edge. The field's positive lead edge will firstly attract the negative lead edge of the electron, so establishing the direction of the electron as exactly opposite to the positive field. The energy of the field now transfers to the electron to back-fill the electron wave behind the lead edge, so producing the electron's momentum. Thus the direction of the electron is exactly opposite (180 degrees) to that of the incoming positive field.

A negative photon will have the same gravitation property as the electron's negative field and so will pull an electron to a wider orbit in an atom before adding its energy to the orbiting wave to secure the new quantum level.

The fact that the electron produces a larger magnetic field than one would expect, may be due to the additional energy of a positive field in space, which exists and touches an electron all the time, including within laboratories when making such measurements.

Some final logic. If we assume that a proton was the first creation in the Big Bang, then a positive field was the first wave in space. That wave could have been the source of the tiny wave that is the structure of an electron. The positive field would cause the electron wave to rotate with an opposite lead edge and so become negative. Thus every proton would have produced an electron, not a positron, and this would have occurred before the positive field added momentum to the electron. If the proton were negative, perhaps the field would be negative and would have created a positron.

But all that discussion – which may be wrong - does not answer the original question: what is energy? It only describes two structures containing energy.

The source of energy

The question of energy is really more about where does it come from? To know what something is, it is necessary to go back to its source. What is the stuff that creates the field and velocity?

Proton Radiation is not the answer as these merely spread the energy around the universe, so the answer is that all energy comes from within the protons. The real energy is whatever is inside a proton that leaks out to create space and gravity, and there must be a lot of it because protons inside Earth have been doing that for four billion years and probably a lot longer before Earth even existed, and there is no sign of it running out.

There must be enough energy inside a proton to supply a house for a week, so if we want cheap energy we just need to find out how to make a proton leak energy faster. That would be a useful project for a university to explore.

We know that inside a proton there are smaller stores of energy called Quarks. Presumably the store is some form of electricity that we may never understand, but the question now becomes; where did all the energy come from to fill all these quarks and form protons? I discuss in the next chapter, my concept that energy is the opposite of 'nothing' and therefore must always exist. The Big Bang was the beginning of its distribution and that is the subject in chapter ten.

So I have not found the answer to my question; what is energy? And perhaps nobody ever will.

But there are some simpler conclusions about energy that I list below.

Energy will always attempt to achieve equilibrium. Particles will move to balance voltage; it is a current. Heat will move to balance temperature; it is photon emission and absorption, or cooling and heating.

The amount of energy leads to a natural progression:-

Low energy creates the charge on two equal and oppositely
    charged particles.

Higher energy and particles will travel with velocity. Radiation of
    electromagnetic waves will occur.

More energy and the waves reach the maximum velocity.

More energy and the wave will grow and oscillate as a wave of
    'black'

More energy and the wave will increase the frequency of
    oscillation in the visible light spectrum.

More energy and the wave will grow larger with higher frequency
    than the particles can support, causing the wave or field to
    disintegrate and form into new particles that divide up the
    energy.

Thus there is a limit to the level of energy a natural particle can carry.

I have produced what I believe is the life cycle of energy at the end of this chapter. In the next chapter I suggest that energy has always existed because it is impossible for it not to exist! It is well known that energy can never be destroyed, but nor can it ever be created. I suggest that energy cannot leave our universe and that Big Bangs have been occurring successively forever – except that there is no such time as 'forever' because time stops at every Big bang.

Protons produce millions of waves a second and have done so for millions of years, so the waves are pretty trivial things but protons are pretty important.

An experiment was carried out in 1932 using an apparatus called a cloud chamber. A very high- energy photon was seen to produce an electron and a positron, which are equivalent but oppositely charged particles. This was believed to prove Einstein's equation $E = mc^2$ because it was thought that a photon of pure energy with no mass was being changed into two particles that had mass. That may be one way of putting it but I prefer to say it shows that energy will produce particles.

The energy in a wave has no polarity because it can transfer between two waves of opposite polarity. It is the source that decides the polarity of the wave, and this is reflected in the structure of the wave, ie the polarity of the leading edge.

So if energy is free by itself it must immediately produce two particles, of opposite polarity, to carry the energy. Now in my view the energy in a wave can equally be considered to be 'mass' because the wave is the device that is momentum.

So in my opinion that experiment actually proved that when excessive additional mass (the electromagnetic wave) is forced to change back into energy the energy has first to form into particles capable of carrying it.

Thus, a) Energy cannot exist by itself, and b) It will always polarise into an equal number and size of positive and negative particles.

An experiment exactly the opposite of the above showed that if two oppositely charged particles collide they are both annihilated and energy is released. The energy then forms two new oppositely charged particles. That seems to confirm that particles are just a store of energy, but these experiments show that in the right circumstances energy can make something out of nothing by producing two opposite particles, therefore energy can convert to particles and become the building blocks of atoms and matter. And energy and matter are the same thing but in a different form and everything in the universe started as energy. The only question in the creation of everything therefore is where did that energy come from? That is the subject of a later section.

Scientists also believe that a neutron in the nucleus of an atom has no charge but does have mass. But everything in the universe is electrical so it is impossible to have any particle that does not have a charge. The neutron must be made of two equal and opposite charges so that they cancel and it is these two charges that produce the electromagnetic wave. In fact the neutron is made of quarks that do have a charge. It may be that because the entire universe is full of positive proton waves, what we call 'zero volts' is actually positive, and so when scientists say a particle has no charge, it really means that the charge is the same as the proton radiation, positive? And a neutron can only have mass if it produces a field, and to do that requires a charge.

The energy released from a nuclear bomb is transferred instantly by photons from electrons, to oxygen and nitrogen atoms in the air that become extremely hot, expand and vaporise. The air or gas pressure and radioactive gamma particles are then pushed outward with enormous force, flattening everything. At no time does 'energy' do anything by itself.

Other ways of looking at $E = mc^2$

Space, and gravity are variations of the same electromagnetic wave, and this wave is mass, so equivalent expressions of $E = mc^2$ could appear in many different forms and I give some extremely crude non-mathematically correct examples below.

Energy x time = momentum x velocity

Energy x time = An electromagnetic wave with velocity x velocity.

High Energy x time = Space x Speed of light.

The Speed of Light

I have explained this briefly in the earlier chapter on time but I will cover briefly it again here.

There are two ways to explain it. One is the mathematical solution of Maxwell's equation for electromagnetic waves in which the magnetic field drives the electric field and that then drives the magnetic field, and so each field drives each other, but a velocity is reached when that driving force cannot move fast enough to drive its opposite field, and that is the limit to velocity.

The other is the same effect but with a different argument. The

constant rate of converting added energy into a field, means that a velocity will be reached when that process cannot be completed in time, so no new field can be produced and that means that the field length is at its maximum, and that means momentum and therefore velocity are at their maximum, and that applies to any field whether a photon or a particle.
.   A radio wave is the same principle. The wave travels away from the transmitter at the fastest rate that it can be created.

Length contraction at high velocity

The theory put forward by Einstein 100 years ago is incorrect, for the same reason that Time Dilation is incorrect, but it seems likely that length will indeed contract, and I believe experiments have been carried out to show it.

A spaceship travelling at the speed of light would have to move the electrons in its atoms into flat orbits like the planet Saturn's rings at right angles to the direction of travel because they cannot have a forward vector as that would require a velocity to be faster than light. Providing further energy would cause the spaceship to melt and disintegrate because the electrons in all the atoms would be forced out of orbit.

(a) A STATIONARY OBJECT

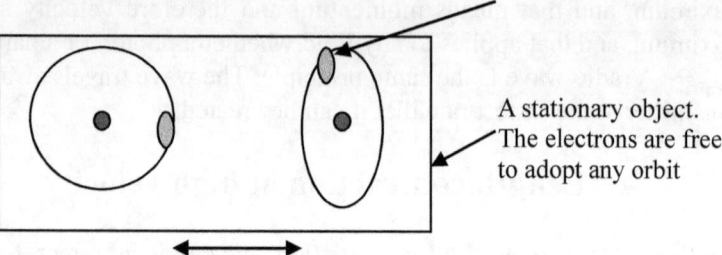

Electron orbiting the nucleus

A stationary object. The electrons are free to adopt any orbit

The atoms are held apart because the electrons repel each other, but are held together because protons attract the electrons forming a balance of forces.

(b) A FAST MOVING OBJECT

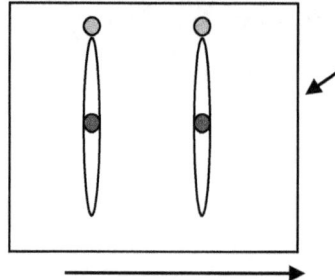

The same object travelling at the speed of light.
The electrons cannot orbit in the direction of travel. They orbit in a vertical plane so the atoms can move closer together. Gravity increases and assists this compression

Direction of travel

Length contraction
Showing why atoms may move closer together at the speed of light, making an object shorter in the direction of travel.
(This is not Special Relativity because that theory is incorrect)

THE LIFE CYCLE OF ENERGY.
From Big Bang to Contraction

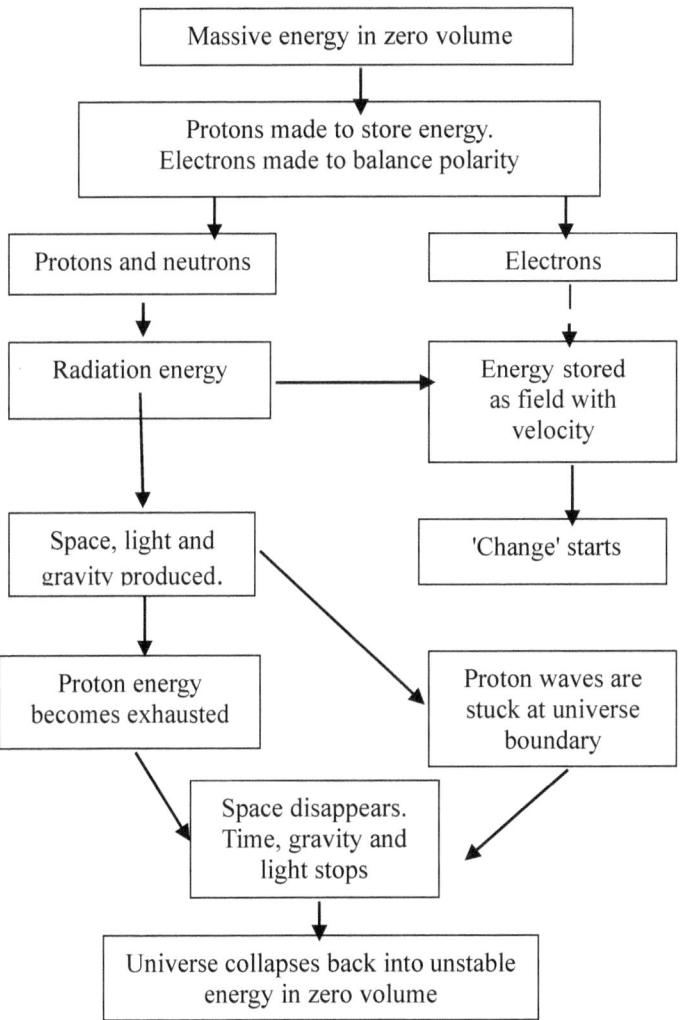

A Summary.
If you have got to this point in the book and have not understood a word of it, as a friend of mine once said about my earlier book, just memorize the following things so that you can really impress your friends.
- Tiny bits of electricity are called particles.
- Big ones form atoms, energy is radiated out of atoms (positive radiation) and travel in waves at the speed of light
- Space is this positive wave of electricity and magnetism produced by the proton radiation. Light is the negative wave
- The process of producing a wave is the cause of resistance to motion. The wave is the concept of 'mass'.
- Visible light is the energy that is dumped from a negative wave when it hits something.
- Time is the man's measurement of change. 'Change' is a change of energy on electrons causing them to move.
- Gravity is the transfer backwards of positive wave momentum to any negative object that is in the way, and creates the concept of 'weight'.
- Every object has its own little universe of space, mass and gravity.

The Basic Laws

I have left this until last because without all the preceding logic it would not make sense. There would seem to be some basic laws,

Law 1.
Energy cannot exist naturally by itself; it exists only in relation to particles.

Law 2.
Everything in the universe is made up of charged particles with waves.

It follows that space, time, light, dark matter, dark energy and gravity are all generated either by charged particles or waves attached to or radiated away from particles.

Law 3.
All particles must be surrounded by an electromagnetic field.
It is a spatial dimension without which they cannot exist. It is space.

Law 4.
$E = m(c^2 + fv^2)$ where E is energy; m is mass, f is frequency and v is the mean transverse velocity of the frequency, c is the speed of light. Put more simply it is $E = mc^2f$. Time is essential for the conversion, so perhaps it should be $E\,t = m\,c^2f$.

Law 5.
Everything in the universe is governed by one constant – 'The rate at which a charged particle is able to convert energy into a wave'. This leads to the maximum velocity and Maxwell's equations.

    There has been a debate about why the laws of physics are exactly right to produce stars, planets and life. For example,
    If gravity were too high the universe would collapse.
    If gravity were too low hydrogen would not merge into clumps and fuse to produce stars.
    These arguments led to M-Theory which proposes that there are many universes with different properties and ours happens to be the one with the right laws. In my view that is nonsense as the laws of physics cannot vary. The speed of light will be the same in every universe. But also I do not think there can be multiple universes because there can be no dimension (space) between them and they would all merge and become one universe.
    Perhaps the correct answer is 'why ask the question as no reason is required'! Physics should be spending its time and money on things that have a commercial benefit and not on solving unnecessary questions.

Chapter Ten

MY THEORY OF CONSERVED RELATIVITY.

This chapter includes the Big Bang, Black Holes and the contraction of the universe and the theories are based on the theories in the earlier chapters. There is no question in my mind that existing theories about Black Holes are incorrect. But first I will discuss relativity.

Relativity

In this book my term 'relativity' is the relationship between two separate electromagnetic fields from two separate sources, or objects (i.e. frames of reference). I will explain that without this interaction, no object can move at all!
    Relativity, when observing an object, is observing the change on each photon wave as it passes our eyes, and each stream is a snapshot of light energy on a photon emission from the object. Because the changes are a stream of millions of photons a second, we see the instants as a continual line, whereas, just as the pixels change in a TV picture, the image we see is actually made up of fast moving dots of photon waves.
    Relativity can only exist between two things if their energy streams overlap. Every object creates its own separate energy 'mini universe' of space through its own emissions, and these are exactly the same as the total universe in which we live. If they did not they could not exist at all, but as all these mini universes overlap, relativity between them is possible.

Conservation of Relativity.

You will need to bend your mind to understand this theory!
    There is a law of Conservation that states that 'energy can neither be created nor destroyed'. It would be simple to say that this law could be extended to state that 'energy within the universe must be conserved', that is; energy cannot leave the universe.
    Putting the theory of Conserved Relativity simply as a law, it is:-

"Nothing can leave the universe because relativity, once created, can never be lost".

But such a statement, whilst true in my opinion, does not explain the reason, or the important consequences.

Firstly, remember that I have said that at our human level, relativity is 'observable change' between two frames of reference, and at the atomic level, it is the overlapping electromagnetic wave between two particles, or the link between the field of one particle and a second particle. They are all the same thing, but the following considers relativity at the atomic level.

I have also said that the proton radiation leaving stars produce the wave that we call space, but may not hit another star and instead may travel at the speed of light forever. A distance will be reached far away from galaxies where the particles have such low density that their waves no longer overlap, and I described this as being the boundary of the universe. Space clearly cannot continue infinitely. I then suggested that the waves may continue forever through 'nothing' as an isolated blip of energy, but would not move because velocity is meaningless without relativity. However, I now believe that a slightly more accurate description is possible as I now explain.

A stream of waves would be extremely short because it is an emission or radiation in packets, so this would mean that at the boundary of the universe, these short waves would quickly lose relativity with each other and become completely isolated.

Looking at it from the perspective of the human level and observable change, velocity without relativity is meaningless. Beyond the boundary of the universe, no matter what the velocity, an object or particle cannot move because there are no reference points. The object is in a place where there is no volume or distance, and without distance, movement is impossible. The 'place' does not exist! The object is creating its own space just for itself, but if there are no other objects or waves in contact and there is no other space, the object cannot move. Yes it has velocity, but no it is not moving! Relativity is more than just 'observable change'; it is also the dimension of volume, or space, and the need to touch another wave.

Consider a case of a proton wave. It becomes stationary immediately it begins to lose relativity with other waves at the boundary of the universe due to the reducing density. Now as more and more waves reach the boundary of space, the boundary will move slowly outwards because the density of waves will allow relativity to extend a little further each time, (and also because the galaxies are expanding outward) but there will always be a sphere of positive waves forming a clear-cut boundary around the universe beyond which is 'nothing'. Remember 'nothing' has no volume, no distance, no contents, no space; it doesn't exist. (It is difficult to get to grips with a place that doesn't exist isn't it!)

The only thing that exists anywhere is our universe. There cannot be two universes because they would immediately become one universe. There cannot be a 'negative universe' or mirror image of our universe. There is nothing outside our universe and because there is no 'distance', anything out there would have to be touching our universe, and that means it is within our relativity and therefore within our universe. The theory of a parallel negative universe is impossible because it would lead to annihilation of both universes immediately.

Thus because proton waves cannot leave, it means relativity must always be conserved, and therefore energy must always remain within the universe. Note that this also means that the universe will always remain in equilibrium. The level of positive charge will always exactly balance the level of negative charge. Equilibrium of charge and energy within the whole body of the universe is conserved. We could think of the universe as being a single gigantic neutrally charged particle.

The theory that two universes were created in the same space, in which universes of matter and anti-matter fought for survival, and our universe won, seems to me to be unnecessarily complicated and is discussed in the next chapter.

One could think that, if there is no 'distance' outside our universe, a spaceship could leave on one side and return instantly on the other side, and so cross the universe in seconds. Equally, one could argue that the proton radiation does leave the universe, but then immediately returns at a random point elsewhere on the rim of the universe, but both of these are impossible. It is impossible to leave the universe because that would require relativity to be lost. A spaceship could move the boundary of the universe because it can produce space, but as soon as relativity begins to be lost, the spaceship will stop travelling. It may still have velocity, but it cannot move. There is no distance to move into in 'nothing'.

Equally one cannot argue that if there are no dimensions outside our universe then the whole of the boundary must curve and touch itself so that there in fact is no boundary. But this would mean space would have to curve in an extremely complex manner, which it cannot do because the waves that form the volume of space have straight momentum and there is no other obvious force in space that can bend such momentum. The concept of 'nothing' is difficult to understand but it must exist.

The Contraction of the Universe.

As more and more proton radiation and photons leave atoms and push the boundary of the universe outward, atoms at the centre of the universe would have less and less energy. Also, as each star burns out through nuclear fusion, they radiate their energy in the form of heat and light, and eventually, all stars would either collapse instantly as Supernova, or go beyond the stage of Black Holes and become debris containing loose electrons with energy equivalent to -273^0C. The protons and neutrons would have no energy at all and so would disappear. Gradually the entire centre of the universe would consist of loose electrons with little energy while the outer boundary would become larger and larger, and more and more positive with masses of proton waves.

The nature of energy is that it always seeks equilibrium so that the level of energy throughout the universe remains broadly the same. If no energy is left in the protons inside the universe, no more radiation can be produced, so the production of space would cease. The relativity between proton radiation streams would quickly become broken inside the universe and that means the dimensions of the universe would disappear instantly. Distances in all directions would very quickly become zero. (But 'very quickly' can mean thousands of years).

The electrons, protons and neutrons left in galaxies within the universe would instantly contract into a dot, followed quickly by the positive radiation and photons at the immensely distant boundary. Positive and negative particles would annihilate each other leaving their waves (energy stored as momentum) to become free energy. Even those galaxies that were new and could still create space would find that the space around them has disappeared and they too become transferred instantly into the dot.

The whole universe would become just a dot of pure energy in zero volume – except that there is no such thing as 'pure energy' – it immediately has to change to take a form, either as potential or kinetic energy or Higgs bosons. But, in effect, the dot is when time resets to zero.

For an instant, the dot would have no gravity, or light, or time, and probably no volume because there are no atoms and no protons pushing out waves and creating space and gravity. Energy would build up in the dot, not as heat or mass because I believe both require oscillating electromagnetic waves, and in a dot that would be impossible. The energy would exist by itself in a form that we do not yet understand, perhaps something like quarks, constrained in zero volume, and that is an unstable situation that must be corrected. Thus something would have to happen, but I am not convinced that there was a bang or any kind of explosion.

At the instant the universe contracted into the dot, time (if we continue to use that word) would stand still. That instant is the nearest thing to 'forever' because there would be no 'change', no measurement of time and no memory of previous time. It is as if there were no previous time and the universe is about to start for the first time. Because everything has stopped, including time and production of space, you could say that the universe does not exist. It is pure energy in zero volume, except pure energy does not exist, it has to be potential or kinetic, which means particles must exist immediately.

Without time or space, existence as we know it is impossible, but of course it still does exist. Thus the universe can be in the dot yet not exist! An external observer would not see the dot because there is no light or movement.

To correct this impossible, and therefore unstable situation, energy would begin to polarise into positive and negative particles. I will discuss this below in the section on the Big Bang, but radiation would then re-start from the bosons and protons to produce space, gravity, atoms and a simple form of time. So change (time) restarts, and that would be the nearest thing to the 'creation'.

The collapse and restart may have happened just once or it may have happened a thousand times, but that becomes irrelevant because there is no history. There is no universal time, and each time the universe collapses, 'change' stops, all history is eliminated so that there never was an earlier time!

Because, in my opinion, the process of the universe is that of a never ending cycle of expansion and contraction, one cannot say that there was a starting point called the Big Bang, as that was just one element of the cycle. Furthermore, I am not sure that understanding the Big Bang adds anything to improve our simple everyday lives. It is just too different! However, I will include a brief comment.

The Big Bang

This is so far removed from school physics that I am not in a position to comment, other than to raise some questions and suggest an alternative solution based on earlier chapters in this book, and I am not sure that finding the answers for the Big Bang is of any value, other than to give satisfaction to scientists. It will not reduce my electricity bill.

The accepted theory for the Big Bang very briefly is,
1. The dot contained protons, neutrons and electrons. It was very small, highly compressed and dense with a mass of trillions of tons, and extremely hot.
2. Vacuum energy produced gravitational waves and ripples in space so producing 'Inflation', meaning massive expansion of the minute universe in a fraction of a second and faster than light.
3. This expansion allowed the energy in the universe to become homogeneous and that would not be possible without such rapid expansion.
4. Then everything cooled due to the expansion of space until the heat was exactly correct for hydrogen fusion and the production of helium and the release of energy.

5. The hot gases were in the form of plasma. This cooled further because energy became spread out by bosons as the volume of the universe increased.
6. This cooling enabled electrons to adhere to the protons, so creating hydrogen atoms. This period has been called 'Recombination'.
7. This produced light for the first time from a homogeneous soup of atoms, but the stretching of space during expansion caused a Doppler or red shift in the wavelength of the light so changing the wavelength to microwaves, and these are the microwaves that have been detected with radio telescopes.

The main reasons this became the accepted theory is because it produced the correct level of helium that exists today, and the microwaves showing the homogeneous universe have been detected.

As I have said, my opinion carries no weight compared to that of scientists, but if my theories discussed in this book are correct, then I have problems with this Big Bang theory as follows.

I do not believe the dot could contain any electromagnetic field and therefore could not have mass or temperature.

In chapter six on space, I suggested that for space to have a dimension it must be made up of a material and that material, at the time of the Big Bang, would have been radiation from protons. It is not possible to produce ripples in 'nothing', therefore if ripples occurred they would have to be in proton radiation. Is it possible to make ripples in an electromagnetic field? (Remember that outside the universe there is no space. No dimension. The place does not exist, and that would have been the situation everywhere outside the original dot of energy).

I mentioned in chapter four on gravity when suggesting that experiments that support General Relativity drew incorrect conclusions. I suggested that gravitational waves may be dense radiation from protons and that field is a source of gravity,

I have a problem with the theory that inflation occurred much faster than light because I do not see how that can be possible. At the beginning of this chapter I suggested that there must be conservation of energy within the universe, that is – a light wave cannot leave the universe. It must come to a stop until other light waves touch it, so producing the relativity of waves for motion. Thus the light, or microwaves that we now see, may have taken much longer to reach us than we think because they did not travel at the maximum velocity. They travelled at the rate at which the universe was expanding. Thus there was much more time than the 400,000 years in which the contents of the universe could become homogeneous, so did inflation really occur? Is the universe a little older than 13.8 billion years?

In chapter three on mass, I suggested that mass took the form of electromagnetic waves and was in fact, the length of the field material. So the only possible mass in the universe at this time was the Higgs Field.

So from the above comments, I do not know whether there were gravitational waves in the radiation from protons causing inflation, or whether the microwaves travelled much slower than the speed of light giving more time for the homogeneous universe to occur.

One thing I do not fully understand is that it is said that we cannot see the light of the Big Bang because it has not reached us yet, so how could the Higgs Field reach us to give us mass? Unless of course the continuing proton radiation is the Higgs Field.

Because of all these things I have trouble agreeing exactly with the accepted theory, and my own opinion is quite different, as follows.

My Suggestion.

- Energy was stored within the zero volume of the dot in a way that we cannot yet appreciate, perhaps as quarks, but there was no mass or temperature.
- There was no bang or explosion.
- Protons, neutrons and electrons were all created at the same time producing hydrogen atoms, because that is the only way to achieve the ratio of one electron to one proton. (If there were free electrons and proton gravity, about 20 electrons would stick to a proton.).
- Radiation from protons gave momentum (mass and velocity) to electrons in the atoms so that they were placed in an allowable

orbit.
- Proton radiation also slowly created the space dimension.
- Such hydrogen atoms had to be created sequentially, in time intervals, and around the periphery of the dot because they required space that did not exist within the dot.
- The proton radiation (Dark Energy) carried' the sequentially produced atoms outward, (the expansion defined by Hubble), so producing a spherical homogeneous zone.
- Proton gravity then pulled some atoms into clumps and eventually into galaxies.

But that is as far as I can go, and there are still some 'what ifs' to all of this!

What if the Big Bang was not the first of its kind, but was the fifth or sixth, following earlier collapses of the universe, as suggested earlier.

Each earlier Bang would change the circumstances and elements available for later Bangs. So it is possible that the starting point in our Big Bang was not raw energy at all, but the dot already contained some of the elements that we think our Big Bang had to create.

As I said at the start of this section – does it matter how the Big Bang worked? Will we gain anything if we ever find the answer? Will my car use less fuel? Isn't it all rather irrelevant?

The most important question is what produced the dot in the first place? Was it God? Or does my theory of a never ending cycle of collapse and Bang make sense? If so, why would God create a universe full of life, only to destroy it all later? Is that a final proof that God does not exist?

I am happy with 'contraction'. 'Creation' is the subject of the next chapter.

.

THE CONTRACTION OF THE UNIVERSE LEADING TO ANOTHER BIG BANG

Contraction.
One Second before the Big Bang

Space creation has stopped. Everything collapses back into a dot including old proton radiation

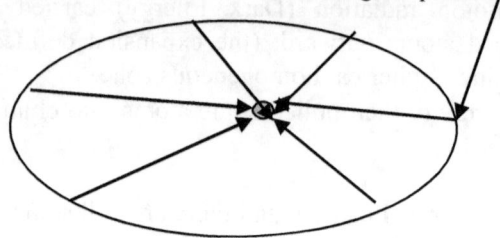

Protons and Neutrons are exhausted and disappear. Space creation has stopped. Space disappears, and everything is pulled into the dot.

The Unstable Dot

The universe has disappeared. The dot is pure energy in zero volume surrounded by 'nothing'. Energy cannot exist in this situation and must polarise. But protons must exist first to make space then 'Bang'.

Expansion
One Second after the Big Bang

Protons, Neutrons and electrons are created together slowly to form hydrogen atoms at a rate controlled by space creation by the protons. This space creation carries the atoms outward. There was no bang.

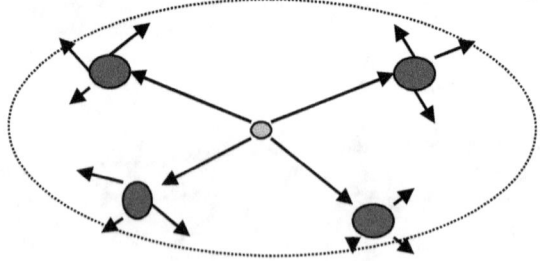

Black Holes

I can see why it is called a 'black hole' because things disappear inside it never to be seen again, but actually it is a black sphere, and my brain cannot visualise how space can be curved around a sphere unless there were trillions of tiny curves no larger than one centimetre. The black hole is always drawn in two dimensions as a hole because to draw it as a sphere is almost impossible.

I have suggested earlier that the reason a Black Hole is black is not for the reasons scientists currently think. Current thinking is that space is so warped around a Black Hole that space bends back and light cannot escape. Or light itself cannot achieve the 'escape velocity' and is trapped inside. This cannot be correct.

I don't believe Black Holes are as mysterious as they have been made out to be. If the process of a dying star were followed through to completion, a Black Hole would seem to be inevitable.

A star is a ball of hydrogen that is so compressed that nuclear fusion takes place changing hydrogen into helium and releasing enormous energy. In every star there are two forces at work. Gravity pulls the atoms inwards and the heat generated in the fusion pushes the atoms outward and there is equilibrium of these two forces for most of a stars life.

When the fusion process has used up most of the hydrogen atoms a star becomes a Cepheid; that is a star that changes its brightness every few days. The reason for the change is that when fusion stops, energy is no longer released and the star cools. The reduction in heat allows gravity to become the dominant force and the star is compressed to a much smaller size. But this compression is enough to restart the fusion process causing heat to be generated producing brightness, and the star expands against the force of gravity. The process is then repeated.

When all the energy in the protons is used up during fusion, the star becomes a Pulsar or Red Dwarf. The star is now mostly comprised of neutrons and the hydrogen is compressed by gravity to a very small size, perhaps only 10 km across. The star has cooled substantially so that the original high- energy bright white light has become low energy at the red end of the spectrum.

The logical next phase is that the star continues to cool until there is insufficient energy even to produce the red end of the spectrum and it appears black. It has become a Black Hole. So one possible conclusion is that light does not emerge from a Black Hole because there is not enough energy in the star to produce light.

But there are many possible reason for a black hole and these are.
- There is not enough energy to produce light.
- The protons are totally exhausted and cannot produce radiation.
- The electrons have stopped orbiting atoms, or are at their ground state and no photon radiation is possible.

The first of these is a possibility as explained above. Fusion will have used up most of the original protons and the remaining fusion may not be adequate to produce enough energy to maintain all the electrons at their highest orbit. Photons will remove energy faster than fusion can produce it. So the proton waves produce gravity, and external light will bend around the hole, but the electrons are too near to their ground state to produce light of their own.

The second cannot be true because Black Holes do have gravity.

It is said that gravity around a Black Hole is immense because the group of stars that make it up have become so squashed that the density of proton radiation, and therefore the force of gravity at this small surface is high. This force could be enough to squash the atoms at the centre so tightly that the electrons orbiting the atoms are crushed and cannot orbit. (The act of crushing probably causes the energy on electrons to be released as Gamma rays). Protons are clearly producing the positive waves because there is gravity, but if the electrons are crushed there can be no energy radiation to form a photon.

If it is true that the electrons are crushed, they cannot oscillate and so there can be no light inside the Black Hole even if the energy level is high. Whilst the centre of a Black Hole may have the equivalent energy of a 'hot' body, that energy cannot come out. Instead, if we measured the temperature at the edge of a Black Hole we would find it to be -273^0C; that of the energy of proton radiation.

Time cannot stop inside a Black Hole if any sort of 'change' is taking place and there will be change because proton radiation is leaving atoms. What we see when we look at a Black Hole are 'snapshots' of black. Because it is black we cannot see 'change', but change will be continuing.

So in my opinion a Black Hole is not much different from any other star, but because it is running out of hydrogen fuel for nuclear fusion, it means the number of protons will eventually become relatively low. This means the production of the radiation that create gravity will reduce, and so it is probable that eventually the gravity will become so low that the Black Hole will become just a cloud of gas under no pressure, and drift harmlessly away into space.

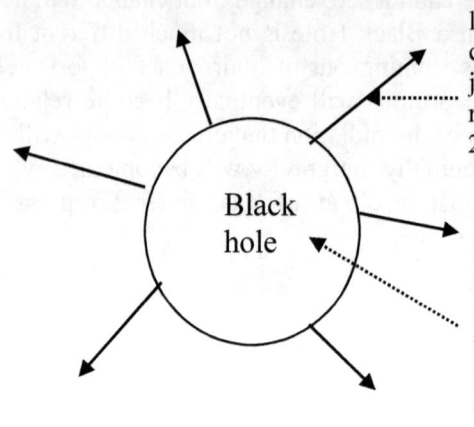

Proton radiation is leaving atoms and creating gravity with just the energy of their radiation equal to – 273 C.

Black hole

Electrons in the star that normally orbit in atoms are pulled out of their orbit by the intense gravity and cannot radiate photons. The star appears black

THE BLACK HOLE

Proton radiation leaves Black Holes creating its gravity, but photons cannot be produced from electrons because electrons are crushed from orbit and have no energy.

Chapter Eleven

MY THEORY OF THE CREATION OF THE UNIVERSE.

Warning – You may have to read this chapter several times to understand the concept!

I do not think that physics can provide the answer to how the universe was created, but pure logic possibly can, and you will need to pour yourself a drink, lie down and read this slowly. It will make every scientists hair curl, but in my opinion it is the logic of the situation. The logic follows on from the previous chapter covering the contraction of the universe and the Big Bang.

There is actually no need for a 'creation' to have ever happened. The situation that currently exists is exactly the situation one would expect to find. The difficulty is that we are misunderstanding the terms 'forever', and 'nothing'.

I will explain the logic in very small steps to help you to follow my thinking.

Nothing.

1. Everything in the universe is made from energy and nothing more.
2. Energy has no dimension. No volume or time. You cannot touch it or see it. It cannot be split into two parts because relativity (my conservation theory) stops it. But we know it exists in the real world.
3. As I have said in earlier chapters, 'Nothing' surrounds our universe, and also has, no volume, or time and you cannot see it or touch it. 'Nothing' cannot be split into two because it has no dimension. Thus 'energy' and 'nothing' are very similar concepts.
4. Does 'nothing' actually exist? There has to be a state of 'nothing', but it is a virtual existence. Because we know there is 'something' there has to be 'nothing'.
5. We know that energy within the universe can neither be created nor destroyed, and everything must have an opposite to achieve

equilibrium. Energy must have an opposite to have enabled its creation.
6. But energy has no polarity; there is no negative energy. The opposite of energy can only be 'no energy', or 'nothing'.
7. So you could argue that the creation of 'energy' must have occurred at the same time as the creation of 'nothing'. But 'nothing' cannot actually have been created; it will have always existed, so we must conclude that energy must have always existed.
8. But the single and only difference between energy and 'nothing' is that energy polarises. It is electric but cannot exist in pure form; it must produce positive and negative charged particles with voltage, rather like a car battery. It is a law of physics. If we could add all the positive and negative charge volts in the universe I think we would find an exact balance. This polarisation could be described as the creation because it is key to producing the universe. Without polarisation, energy would probably annihilate with 'nothing'.
9. Energy delivers a force because it has become out of equilibrium and is trying to return the positives and negatives back to equilibrium; its natural state, which would allow it to neutralise with 'nothing'. It is the attraction and repulsion of these charged particles that produces all the forces and dimensions in the universe starting with velocity, then space, mass, time, gravity, inertia, momentum, atoms, trees, flowers, water and humans. Energy has now created 'something' we can see and which, in our normal perception of things, is the opposite of 'nothing'.
10. We can see, feel or detect all of these. They are dimensions and matter. It is the universe in the real world.
11. So the opposite of 'nothing' in the virtual world, is 'energy' in the real world. Energy polarises and becomes 'the universe', so that the opposite of 'nothing' is 'the universe'. But we know the energy in the universe has a specific level. How can that be explained?
12. The size or value of energy must be a variable depending on circumstances. 'Nothing' is in limitless supply. Energy must always try to balance this limitless supply. It is the basis of equilibrium.
13. Energy produces the dimension of the universe. 'Nothing' is the

exact opposite of dimension, but must grow to surround energy. A spherical ring of positive photons separates the two. (My theory).
14. But, in my theory, each time the universe contracts to a dot the barrier is removed because all the photons smash into the dot and form protons. So for a short time in each contracting and expanding cycle there is no barrier between energy and 'nothing'.
15. During this short time, energy will seek to balance with 'nothing', but it is pointless to speculate further on how size is decided. It may be the speed of contraction, or the size of the dimension that contracts, etc. It is possible that energy increases with every contraction, but that would suggest there was a time when there was no energy. We may never know.
16. 'Nothing' is not a force but it clearly can undo everything that energy has done by virtue of its zero dimension. There are two reasons to justify this. a) 'Nothing' has the ability to stop photons leaving the universe unless there is another photon e. m. wave to secure the energy to the universe, i. e. energy cannot split into two. b) When all the energy in the protons is used up and space is no longer produced, all dimensions disappear. One might expect stars to remain exactly where they are, but that is impossible because there are no dimensions. Space has disappeared. So everything in the universe contracts to a dot. It means that in one instant, 'nothing' has the power to move all the boundary proton radiation and photons with their energy back to the central dot. It can undo in an instant what energy has taken about 50 billion years to do. And, because there is no distance, no velocity is involved.
17. That is a further reason to suggest that energy and 'nothing' are exact opposites and why 'nothing' must be balanced by energy.
18. I have suggested above that 'nothing' and energy have both probably existed forever, but 'forever' is an impossible term.

Forever.

1. 'Forever' is similar terminology to 'infinite'. Both are mathematical terms that in real life are impossible. Only a mathematician will think they have a real possibility.
2. But the universe does not need 'time', it has continual change that produces our perception of time, so what we mean by 'forever' is; will change continue and never stop? If change ever stops, then we can say time stops and there is no 'forever'.
3. As long as there is space and photons to move energy around, change will continue.
4. But the universe contracts and expands as described in the previous chapter, and whilst it is contracted into a 'dot', 'change' has stopped, our concept of time has stopped and space is zero.
5. If there is no space and nothing can change, something cannot be said to exist. Conversely, therefore, existence begins when change (time) and space begin. There is no 'forever'.
6. If we use our own perception of time for simplicity, then time goes in loops. Time re-starts at each big bang, and continues until contraction to a dot, then stops, because change stops. A completely new and unrelated time loop re-starts at the next big bang and completely new change begins.
7. The loops are unrelated. There was no 'first loop'. There is no past. There is only 'now'. The perception of the past exists only because we have a memory and, by association with visible daily changes, causes us to imagine that there must have been a past before the Big Bang.
8. But if time stops at a dot, our understanding of evolution, or 'change' has stopped. The universe is 'on hold'. But how long is 'on hold'? If time has stopped there is no measurement of time so it may always have been stopped! There is no history of whether time has just stopped, or whether it has never started!
9. Whether time has just started for the first time, or it has stopped just once or a hundred times previously, all becomes irrelevant. The nearest thing to 'creation' is the fact that time and change starts.

10. So we cannot really base our solution to the creation on time, we should base it on changes from one state to another, or on cycles of expansion and contraction, during which, time as we measure it, stops.

Another way of attempting to explain this is to consider that you cannot point your finger at 'nothing' because by definition, 'nothing' does not exist. We cannot easily imagine or locate a place that has no dimension but as soon as something such as a universe exists you can point your finger at 'nothing' because it now has a location. It goes all the way around the universe but has no thickness or dimension as it is just a state.

Energy is the same. It is just a state. It has no force, power or dimension. The force of energy exists only because it polarises into two opposite types that must repel each other in order for energy to remain in existence and its existence is necessary because its state has to be different from that of 'nothing' or 'nothing' cannot have a state either, and that is impossible!

This polarisation happens to be electric. Everything in the universe is electric and because this power exists, it must have always existed, and always will exist. Electricity is a law of physics and laws do not change. But time has not always existed, nor has space. It is time and space that are the elements of creation and the proof of existence.

I have attempted to summarise all of the above into four simple laws as follows:-

The Laws of Creation.

1. A state of energy must always exist because a state of 'nothing' must always exist.
2. Energy and 'nothing' are always in equilibrium.
3. Change (That causes us to invent time) stops on each occasion when the universe contracts, and re-starts when new protons and electrons are produced at the centre of the contraction.
4. A new creation can best be said to occur on each occasion that change (And time) re-starts.
It is reasonable to ask the question 'Why isn't there a Big bang

every fifteen minutes'? I think the answer is simply that existing material has to accumulate at the centre to form a large enough amount of energy for a Bang, and so the universe must go through its complete cycle before a new bang can occur. It supports the view that no new energy is being created, just the same energy going through a cycle and being re-used, and so no new energy caused the first Big Bang.

There are also philosophical questions such as 'If God created the universe, what was he doing before that? He had an infinite amount of time to do nothing at all'! That does not sound very probable.

The bottom line question is why didn't the Big Bang occur immediately and the answer can only be that it did. Therefore the only answer is time itself. When time started, the universe started, and that is what I have described earlier.

The approach of scientists

Scientists have been looking for the answer to creation in anti-matter, but that cannot be right because,
   a) Anti-matter is just an alternative configuration of energy. All matter requires energy, so a second universe comprised of anti-matter would require double the amount of energy to have to be explained.
   b) There cannot be two universes because, according to my theory, outside our universe there is 'nothing' and nothing has no dimension: no distance in any direction. Two universes would have to touch and become one. Energy would follow the same rule. There cannot be any energy outside our universe. If my theory is wrong, then to have two universes, space would have to go on to infinity, and infinity is a meaningless term found only in maths.
   c) The laws of physics can start only after energy exists. Energy must come first.

The concept that 'nothing' somehow becomes 'something' is invalid. No matter how long we search for an answer, it is impossible. The answer is you must have both at the same time. You cannot have 'nothing' unless you have 'something'. They are in essential equilibrium.

And nothing can be any size because it has no dimension. The size of the universe will vary from a tiny dot, to a sphere possibly a thousand times larger than it is now, and 'nothing' has a limitless supply; it will always surround it. My theory of conservation of relativity provides a wall that separates 'something' from 'nothing', and no matter how much energy is inside the wall, it cannot leave.

Some may want to argue that there must have been a third state that contained both something and nothing somehow intertwined, but that would be stretching things and it is best to say that there has always been something (our universe), and nothing (the zero distance around us).

Creation therefore re-starts whenever change (or time) re-starts, and there is no other possible configuration of things than 'something' and 'nothing'.

Now all of that is just my theory. Only time will tell whether it is correct, but I would rather believe my theory, than imagine somehow that there are two universes, one all positive, and another, a mirror image, all negative. In my theory, positives and negatives exist together in equilibrium in the same universe; these are produced by energy, and all the energy is also balanced in equilibrium by nothing.

Is the Universe Shaped like a Doughnut?

I have said in the chapter on space that the shape and current size of the universe is the shape of all the galaxies plus the distance proton waves can travel before their density is so low that they lose relativity with each other, but there are other theories, based on maths and I believe Einstein's General Relativity, that the universe is shaped like a doughnut, or bagel. This is not a theory I agree with.

It is possible that because the Big Bang pushed matter outwards, that after 14 billion years the centre around the origin might be expected to be empty. If the density of proton waves and photons were too low at the centre to retain relativity, and if there is a law that 'relativity must be conserved' as suggested earlier, then the particles would form an internal rim or sphere, with a hole of 'nothing' at the centre of origin, and this might seem to confirm such a theory. But 'nothing' has no dimension. It means the front, back and sides of the internal empty sphere must all join together and the central 'hole' then becomes a hole of zero volume, just as the outside of the universe is a place of zero volume, i.e. it does not exist. It is hard to imagine a volume that does not exist, but I think that is the scenario that we would have to conclude, but it is not one to which I agree.

The Accelerating Expansion of the Universe.

Einstein believed that the universe was static; that it was neither expanding nor contracting, but unfortunately his own mathematical calculations showed that the universe would eventually contract. Because he did not like that answer he inserted a 'cosmological constant' into his equation that held the universe static. This constant was the equivalent of an unknown and unidentified force that seemed to be working against gravity, pushing the galaxies outwards.

Many years later, Hubble made observations of the velocity of stars using the Doppler Shift technique in which the wavelength of light changes depending whether the star is moving towards or away from the observer. To everyone's amazement, this showed that the stars furthest away were moving away faster than the nearer stars in a relationship proportional to their distance from the Earth, i.e. the universe was expanding. This same technique allowed Hubble to establish how far any star was away from Earth and also, by extrapolating all movements of stars backwards, fellow scientists realised that they all originated from a single point; the location of the Big Bang. This process also enabled them to establish the age of the universe.

I discussed this expansion in an earlier chapter where I explained dark energy and dark matter. The term Dark Energy is the name given by scientists to a mysterious and unidentified force that seemed to be pushing galaxies apart, rather like the 'cosmological constant' in Einstein's equation. I suggested that this energy is the force of proton radiation elements repelling each other so that the space dimension is increasing and causing galaxies to move outwards.

Scientists seem to accept that this is what is happening.

Differences in Theories.

We all agree that in the beginning there was a Big Bang, but from that point onwards my theory of the universe differs from established scientific opinion.

- Scientists say light does not leave a Black Hole because gravity is so strong it curves space backwards.

In my theory there is no 'curved space'. Gravity is produced by the protons in a Black Hole but the electrons in the atoms inside the Black Hole are pulled down so strongly by gravity that they are pulled out of orbit within their atoms, so the emission of energy from electrons to form photons cannot be achieved.

- Scientists believe that matter has a maximum velocity because all energy changes to infinite mass with no energy producing velocity.

My opinion is that matter and light have exactly the same cause. The energy conversion process of a particle produces a constant rate at which an electromagnetic wave can be produced, including that of a photon, but as the velocity increases, the process is unable to complete the wave before the velocity has moved it away, so both the 'space' dimension and momentum that it must have to be able to accelerate cannot be produced. So there is a velocity at which the conversion of energy into space and linear momentum cannot happen fast enough to allow further acceleration. Instead the energy increases the frequency of the transverse wave. It is the time required to produce the wave that causes the resistance to motion. Mass is the end product, not the cause of resistance.

- Scientists say the universe possibly goes on forever.

I believe that there is a specific boundary where the elements of many proton waves lose contact, and by doing that, the dimension of the universe becomes lost. Beyond that point there is absolutely nothing. No dimension. It is a place that does not exist! And because of this there cannot be 'multiple universes' because the two would join together and form one universe.

- Scientists do not know whether the universe will continue to expand, or has sufficient gravity within it to cause it to contract.

My solution is that the universe will contract but it will not be due to gravity, it will be because the dimension of space will disappear when the energy in the nucleus of every atom expires – as it must. The dimension allowing galaxies to be separated will disappear so that they, and all their energy-less particles, instantly converge into a dot of zero volume. The energy that has accumulated at the universe boundary in the form of photons and proton radiation also collapses into the dot and a new 'Bang' occurs.

- Scientists say time may have existed forever and may continue forever.

My opinion is that there is no time dimension in the universe and time is just man's invention for measuring 'change'. Change is simply due to energy transferring from particle to particle so causing motion. The rate of change is affected by a particle's current velocity and so at high velocity, 'change' will slow. Scientists have interpret that 'slowing of change' as slowing of time', but a slowing clock is nothing to do with time. The energy comes from stars and is carried by photons to electrons, but much of this will be carried away to the edge of the universe and lost. Also, as above for space, the energy producing the proton radiation from atoms must eventually become exhausted, so space will disappear. When protons are exhausted, change stops and space disappears; so time stops but will restart at the next Big Bang. There is no such time as 'forever'.

- Scientists are unsure of the origin of space, time and gravity.

I am certain that every single atom produces its own space and gravity from proton radiation so that every atom, and therefore every object made of atoms, from flowers and trees to stars and galaxies, exist in their own separate micro universes.

So you can see that just from those examples there is quite a widespread difference between my thinking and conventional scientific thinking.

It was not my intention to cover such a wide range of topics and my approach was somewhat unique. Instead of reading up on all the work done by scientists to date I ignored all their work except for Einstein's $E = mc^2$ and Rutherford's atom. I started with a blank sheet of paper and then used pure logic to reach my conclusions. No maths. No experiments. If I had read earlier scientific papers before starting this project I would probably have followed the same doubtful assumptions, and this book would not exist.

Modern Physics

The important thing in physics is to get the simple basics right; mass, gravity, time and space, and that is the purpose of this book. If that is done, not much else really matters, yet three of these have been misunderstood, and if you believe in General Relativity, all four are misunderstood.

Physics has moved away from classical simple physics and into Quantum mechanics, Particle Physics and Theoretical Physics. These are interesting and complex but I am not convinced that they help in understanding the universe, and instead they add an extra layer of complexity. If the basics are correct, much of these mathematical concepts are unnecessary.

The work of Maxwell is brilliant and I touched on quantum mechanics, but I do not find it necessary to know exactly where a particle is within its wave, it is only necessary to know that it has to be in the wave, and that the whole structure has momentum because the wave or field has mass, and that will keep it going in a straight line. So the Uncertainty Principle is only briefly mentioned in this book.

And I do not support the weird concept that an electron has a number of possible paths it could follow, and that some of these cancel each other out, leaving a strong probability of which path it will follow. In my view the structure of an electron and wave, has momentum, and therefore there is only one possible path it can follow – straight, and in terms of understanding the universe, that is all we need to know. (Keep it simple!).

There are some other things scientists believe that I disagree with.

'Gravity' does not bend light waves. I believe it is the outgoing proton radiation that does this, acting on undefined tiny particles within the wave. Gravity of Earth is a big force even if the bits that produce it are extremely weak, so the source of gravity is likely to be from particles or waves that are in abundance and that we already know about rather than something so small and so rare that we have to look for it in an accelerator in the form of new and so far, undiscovered particles, such as the graviton in the standard model.

And then there is all the other stuff about multiple dimensions and parallel universes. Time will tell whether they are right, but I am not even prepared to think about them.

So has modern physics really achieved anything? I just started with a blank sheet of paper, school classical physics and a lot of logical common sense! But perhaps everything in this book is wrong? I do not think so or I would not have gone to such lengths because I am retired from 'work'.

Chapter Twelve

COMMERCIAL OPPORTUNITIES OF GRAVITY

How to Make Objects Weightless.

I see this as an opportunity for someone who is willing to make the investment, and I am fairly sure that little energy is required to stop gravity and so the return in the form of cheap energy is significant.

We can focus light waves using a concave mirror. We can focus microwaves and TV waves using a dish, but we cannot focus proton waves because of their very long wavelength. But there are two reasons why I believe it is possible to make things weightless,
1. Something that happened in the Bermuda Triangle
2. It was achieved by accident in 1998 in Germany by Peter Bettels.

A pilot flying a small plane from The Bahamas to Fort Lauderdale had to fly through a massive electric storm. He reported becoming weightless and that he was suddenly over Miami. He had flown 100 miles in three minutes – a speed that is impossible in his small plane.

So either the bolts of lightning, or the negative electric charge of the clouds seems to have cancelled both gravity and space. It had somehow eliminated the proton radiation.

Peter Bettels had an apparatus in 1998 that used high voltage DC to force electrons to travel very fast through a super-conducting coil whilst rotating the coil very fast to give electrons still more speed, and levitation was achieved, but he has been unable to repeat it. If his result is compared to that of the Bermuda Triangle, then either his apparatus produced a massive negative charge in his laboratory, or his apparatus had an unknown temporary open circuit that allowed gravity to be absorbed?

I show a similar apparatus to the one Peter Bettels used, on the next page which I think may be the real solution to blocking gravity. I believe that the solution is simply to replicate what happens when photons are absorbed by orbiting electrons in atoms. If enough electrons pass across the vertical proton radiation, and these electrons have very low energy, then they will absorb the radiation and then radiate it away.

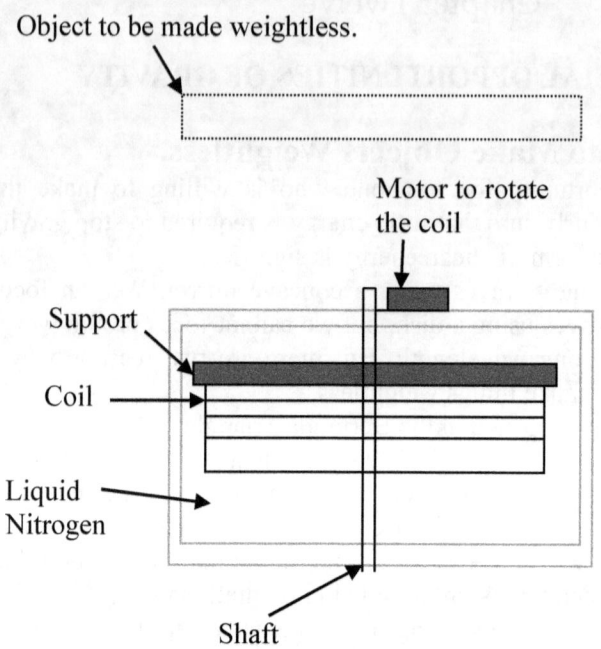

A Possible Gravity Removing Device

A circular coil is rotated rapidly in either direction within a bath of liquid nitrogen. The cold low energy electrons absorb the vertically rising proton radiation and the rotation causes the electrons to radiate it away so that no radiation reaches the object to be made weightless.
This simple design may only reduce gravity by a few percent. To eliminate gravity completely may require a coil several feet high so that there are so many free electrons that no proton waves can pass through.

Is Gravity a Possible Source of Energy?

If gravity can be stopped then it is obviously a source of energy and the best solution to harness the energy is the 'water-wheel concept'.

If you take a Ferris wheel, like the London Eye, and put a gravity reducer under one half, that half will have reduced, or zero weight, whilst the other half remains heavy. The wheel will rotate like a water wheel, faster and faster.

Physicists will say that this is 'perpetual motion' and is impossible because of the laws of conservation, i.e. energy cannot be created or destroyed.

But if my theory of gravity is accepted then it follows that energy is being radiated from every atom in the planet Earth in vertical straight lines, and is being wasted as it travels away into space.

The sceptics will say that the energy added to stop gravity must be greater than can be generated by the device. Only experiments will show who is right.

We have succeeded in extracting energy from light waves via solar panels and so there should be no reason why we cannot do the same with gravity waves. They are similar waves, just different densities and wavelengths.

Other possibilities for extracting energy.

Perhaps proton waves can produce electrical energy if you simply turn a satellite dish to face the ground? Seems unlikely! The wave is too long.

The proton radiation has a magnetic element and when an electrical conductor crosses a magnetic field, a current is generated. So if a copper wire is caused to cross the path of the proton waves rising vertically, a current should be induced in the wire.

Whether the current produced is greater than that required to move the wire can only be determined by experiment, but I believe there will be a small gain in energy and it may be sufficient to keep a 12 volt battery fully charged.

Chapter Thirteen

SCIENCE FICTION:- SUPERNATURAL? IMAGINATION? OR TECHNICALLY POSSIBILE?

My theories in the front of this book are about all the processes of physics that make up the universe, and the single most important theory is that everything, including rocks, trees, animals and ourselves, are all sending out electromagnetic fields at the rate of billions a second.

The question then becomes 'so what'! What does this mean in the real world? Does it shed any new light on other mysteries that we have been grappling with to find solutions? That is the basis of this chapter.

Travelling Faster that Light

Einstein has already proved that nothing can go faster than the speed of light and I hope my theories in this book have helped to explain why this is so. If a spaceship travelled even at half the speed of light it would become shorter, get very hot and eventually vaporise. So does that mean 'warp factor 9' is impossible?

My mind is open on this because I can see possibilities and the Bermuda Triangle story sounds reasonable. I do not think 'worm holes' occur naturally in the universe, but I think it may be possible to create something similar. If so, a spaceship could travel a billion miles by just moving a few feet. That's fast!

The possibility is that if you cancel 'space' you are cancelling 'distance'. Space, which is positive waves, can, in theory, be blocked by firing negative 'electron torpedoes' in front of the spaceship. Immense energy would be required to produce a tunnel of solid and continuous electrons the size of the spaceship that travels forward at the speed of light. These electrons would, in a fraction of a second, block the positive space-making proton radiation in front of the ship so that the space in front becomes 'nothing'. But it would need to be extremely powerful.

But if we can generate such powerful electrons, then, following my earlier law that says 'relativity must always exist within the universe', so that 'nothing' is an impossible state inside the universe, then if space is blocked out in front of a spaceship, the front of 'nothing' becomes immediately joined to the back of 'nothing', and the distance in between has disappeared.

A twenty second burst of electrons would destroy nearly four million miles and if the spaceship travelled at just one quarter of the speed of light then it would travel one million miles in a second. It would in fact re-enter the universe one million miles further forward than where it left. That is 'warp factor 5'!

In effect, rather than moving faster, the distance in front has been shortened, so travel is safe and there could be no damage to the spaceship. It is a 'worm hole' that has no hole in the same way as I described the 'bagel universe theory' as a bagel with no hole!

The problem is still that the electrons cannot move faster than light, so if you wanted to travel five light years, the tunnel would have to be created and held in place with massive energy for five years. Then, when the tunnel is completed, the spacecraft can make its five light year journey in just a few minutes.

My doubt about this is whether a spaceship could move at all if a beam of electrons has blocked the waves that are space, but then perhaps if the electron beam were placed a few feet in front of the ship, it could work.

Other possibilities for travelling faster than light are,
- a) To understand the process by which electrons produce an electromagnetic field so that ways can be found to make it faster. That would mean that spaceships could travel faster.
- b) To produce man-made space by generating an electromagnetic field in front of the space ship.

If a spaceship was surrounded by a strong but adjustable positive field it may be possible to hover over Earth and move upwards at a velocity but probably not near the speed of light. If the spaceship was constructed of positive particles – which is probably impossible – then achieving the speed of light does become possible, but I don't think humans can stand the increase in mass.

Will we ever meet Aliens from another planet?

In my opinion there is no doubt that there are millions of other planets in the universe and a high percentage of these will support life. A smaller percentage will be able to support intelligent life, but that is not the problem. Firstly, such planets are hundreds of light years away, and even if this intelligent life could achieve the impossible and travel faster than light to reach us, there is still a problem.

The problem is whether any life can evolve fast enough to be sufficiently intelligent to escape from its own planet before a major natural catastrophe eliminates it. If we consider that the universe has existed for 14 billion years, and the planet Earth has existed for 4 billion years, then a lot of time passed before the planet Earth even existed. Then about 3.9999 billion more years passed before life evolved to a level of some intelligence, having been almost eliminated by several catastrophes in the meantime. So in all those billions of years, man has existed for only a tiny period of time.

On some planets intelligent life would have started perhaps two billion years earlier than us. On other planets intelligent life may be possible, but needs another billion years to evolve to achieve it. So the problem is one of evolutionary time-scales and catastrophes. The likelihood of two planets developing intelligent life during the same 5000 years, and not having a catastrophe to eliminate it is remote. But there is still the problem of the immense distances. It would take two or three generations of any intelligent life-form just to get to some of the planets.

The good news, in my opinion, is that if we ever did meet other intelligent life, they would, (a) Be nice people, and (b) Would look very similar to ourselves but not necessarily the same size. Those views are simply based on the fact that all evolution must lead to the same end point (man, or a later version of man), and any life that has evolved far enough to have achieved space travel will understand that war achieves nothing.

If we are to preserve our own civilisation a large number of people must colonise another planet within perhaps the next 500 years. If we fail, then a natural catastrophe will send us all the way of the dinosaur.

Chapter Fourteen

POSSIBLE EXPLANATIONS FOR ESP

Telepathy

Man has already evolved from being an ape, and it is reasonable to assume that evolution will continue. If so, what form would it take?

Newspapers reported a scientific discovery in 2004 that sending electromagnetic waves into a person's brain seems to provide a cure for depression, Parkinson's disease and migraines. That newspaper report indicates to me that e. m. waves do influence the brain, but whilst all objects produce waves, the brain is special because it produces its own electrical impulses when thinking. These impulses can eventually create confusion, over excitement and headaches that the externally applied e. m. waves somehow correct.

It therefore seems possible that the brain could place its electrical thoughts onto the cell's normal atomic particle wave emissions (proton radiation) in much the same way as radio signals are placed on a long carrier wave.

Such evolution would only be a matter of learning how to use the 'thinking' impulses of the brain to modulate the particle waves produced by the atoms of the brain. The receiver would do the same in reverse. And, as discussed above, these waves are possibly within the infra-red or microwave spectrum and invisible to humans.

Evolution is a large and fascinating subject of its own. If you ask 'Why did man lose his tail'? The answer is either because it became a burden leading to an early death, or because he no longer needed it. Contrary to Darwin's theory, I favour the latter, i. e. if you don't use something; you lose it. Now turn that around and it becomes; if you use it, you keep it. That is the basis of a theory of 'use-driven evolution' that Darwin believed but modern scientists do not. I have discussed this in another book on evolution, genes and the prime cause of cancer, called *'Man Designed Himself'*, which is available now from on-line retailers.

It would seem to me that we could evolve simply by continually using a muscle or sense no matter whether it helped survival or not. If that is so, then the more we use our brain to try to communicate telepathically, the more we will be able to do it. It is just a matter of time and genetic reproduction. My guess is that in 300 years, telepathy will be seen as normal, and visiting aliens would all have the capability.

Telepathy would have to use a negatively charged signal and the nearest signal that can be created in a laboratory is that of fibre optics, where a signal is originated by light. Perhaps one day this technology could be used to try to send messages directly into the brain. If we do have telepathic ability perhaps we could experiment to determine its wavelength. Then we could build a machine to receive that wavelength, convert it to a current of electrons, place it on a TV screen, and see people's thoughts! We may even learn how to talk telepathically to dogs, or think in front of a computer, and it will respond!

Astrology

It is known that all human personality is based on a permutation of just four elements, leading to a total of sixteen personality types. In my case my elements are 'planning', 'head', 'facts', 'introvert', leading to me being a 'realist' type. I cannot change what I am. Each element has two possibilities; like a positive and a negative, and the alternatives, as I recall, are 'spontaneous', 'heart', 'wishes', and 'extrovert'. My personality is similar to 'Spock' in Star Trek, which used to amuse my wife. I have zero emotions until an exceptionally pretty girl says, "Hello, how are you". Then my brain turns to jelly. In the good old days, when there were people called 'secretaries', I always made the mistake of recruiting the prettiest girl that applied, so my brain became jelly every day. It is because I am a Leo!

Some would argue that personality is inherited through the genes and for many of the characteristics of a person this may be true, but I am suggesting that there is another logical explanation.

I have a theory on evolution that the energy attached to the waves coming from the sun, moon and planets either as photon energy or as electron radiation, enter the body either through the eyes at birth (if due to photons), or through the skull at birth (if electrons), and these wave are able to form certain neurons into specific wavelength and energy level receivers. These are the wavelength of energy that such a neuron would then release when stimulated by a situation.

If these waves from the planets produce the four elements and sixteen permutations, then personality could indeed be set at the date of birth, or the date when a baby first opens its eyes. It would be possible to choose the personality of your baby by blindfolding it until the month of your preferred star sign! Equally, if we knew which wavelengths produced each of the four elements that combine to produce a personality type, we could produce light bulbs with those wavelengths. Then we could visit a local pharmacy and buy a set of 'Leo personality light bulbs' to produce the personality that we desired in our new baby!

Some Sports Trivia.

Lastly, as this is the end of the book, here is an interesting thought for sports fans to discuss in the pub.

If an athlete wants to break the world record for the 100 metres sprint, does it matter in which direction the track is orientated?

The surface of the planet Earth is travelling from west to east at the equator at a speed of about 1000 mph, and slightly slower as you go north. If the athlete runs in the same direction, from west to east, the ground under his stride moves in the same direction so that he covers less ground each time and the 100 metres is longer than 100 metres! If the athlete runs the opposite way, from east to west, the ground moves backwards under each stride so that the distance to run becomes less. If the athlete runs north to south, or south to north, the movement of the ground has no direct effect on the distance, but the athlete would have to use energy to remain vertical because the feet would be continually pulled sideways whereas the body would not. This would mean that there is less energy available for the sprint.

Some of this effect is of course cancelled by momentum. If the athlete runs in the same direction as the earth's rotation, the athlete starts with momentum even before leaving the blocks.

However, consider one further factor. If the athlete chooses to run west to east in the same direction as the Earth's rotation then the absolute velocity achieved is much higher causing inertia and mass to be much higher, so that extra energy is required! Running west to east and the converse is true – it should be much easier. But does 'relativity' come into this!

Chapter Fifteen
MY PROCESS

In 2003 I read Stephen Hawking's book 'The Universe in a Nutshell' and everything was fine until I got to page 42 where it talked about ground state fluctuations, vacuum energy and branes. I had trouble believing that such trivial energy could shape the universe the way it is.

So later that year I started with a blank sheet of paper in my back garden and began with the basics such as what is mass? Why is there a maximum velocity? I simply wanted to work out what gravity and space were, and it seemed more useful and more interesting than doing a crossword puzzle.

I obviously knew of $E = mc^2$, but my question was why does mass increase with velocity? The only thing that happens when a particle is accelerated is it produces a wave, so the wave must be mass. And if the universe is an empty void of nothing, the wave of a particle must also be the space dimension it requires to be able to move. Also at this time I knew about Spontaneous Emission being the process of producing photons or light waves, but I had trouble with the word 'spontaneous' because it implies that it may not always happen, and that cannot be right! There had to be a more consistent process that ensured photons were always released.

Then I read somewhere about Higgs Bosons. Initially I thought that they were radiated from the proton as it decayed, so that seemed to be the solution to photons. The bosons being positive because they come from a positive proton, must collect energy from the orbiting electrons to form a photon. So photons were positive and contained a particle, and that theory formed the basis of my first book. But I was wrong!

It was not until 2011 when CERN was hunting for the Higgs boson that I realised the Higgs bosons were just thought to be the cause of the field that filled space and they were created by the Big Bang. So, as I was certain a positive wave must come from a proton I reached the decision that it was simply radiation' from proton decay that may not be the Higgs Boson, and that is the basis of this new book.

I don't claim to understand any particle physics so I don't understand the purpose in confirming that the Standard Model is correct, (and I still don't, because I did not need it). If you know a proton must decay and radiate its energy away, and you know a proton is positive so the radiation must be positive, why is it necessary to know precisely how it causes the radiation? Or how it fits into a grand plan for all particles and waves? Surely there must be better things to do. So I ignored such details and stuck with the school physics that I know.

The same is the case with mathematics. In my opinion mathematics is good for proving whether a theory is correct, but it is dangerous if it is used to produce a theory, because maths is a virtual subject and you will get silly theories such as multiple dimension and multiple universes that cannot be correct in the real world. People who are brilliant at mathematics seem to lose the common sense that is required for practical physics. Their brains are wired differently.

Whilst mathematics is essential, I believe it should be a 'support' function, rather like a company's Finance and Accounts departments. The CEO in a company is the driver of the business and understands the market and the products. The Finance Director does the maths to ensure the company can afford the investment and will make a profit. The head of Physics should be a practical person who understands the universe in terms of practical physics, and has done practical things himself at home, such as renew a kitchen or bathroom. He knows what is possible and what is 'silly'. His leadership theories are then reviewed by the team of mathematicians for feasibility. If that were done, all theories would be properly validated, and would not be 'silly'.

Where maths would be useful is in Biochemistry. The human body operates using the sun's energy. Strands of DNA are levels of energy. So the best way of curing such things as cancer, Parkinson's and Alzheimer's is to understand where the energy is going wrong using maths.

People like me who propose theories that fall outside accepted scientific understanding are rudely called 'Nutters' so it is hard to get new theories accepted by the profession. However even the average man in the street would have the sense to know that something is wrong with the theory of multiple dimensions, but we assume the professionals are brilliant people. I am not so sure now, and worry that, in their academic isolation, they have lost touch with common sense and perhaps some of them are the real 'Nutters'? **The number of errors I have found is disturbing for a highly respected profession.** Why have they not realised in 100 years, that much of Einstein's and Lorentz's work is clearly invalid and that their understanding of mass is wrong?

I am a practical engineer, not an academic. I have built an in-ground swimming pool, (including using a power digger) a Jacuzzi, a boat, a conservatory and modernised several kitchens, and it is these practical ventures that allows the brain to decide whether theories are possible or silly, and I suggest that all academics get out of their 'ivory towers' and spend considerable time doing 'real world' practical things.

My technique is to look at processes, look for the simplest solution and think 'outside the box' of scientific answers. I knew that because scientists had not resolved gravity, even 100 years after Einstein's General Relativity, something must be wrong. I fairly quickly resolved what space and mass were and that led to the realisation that the resistance to motion must be the process of producing mass, not mass itself. That had to be the case because it was the only way that could lead to the same maximum velocity for every type of particle. It was a standard common electrical process and there was clearly a velocity at which the process could not produce the result needed fast enough. So acceleration stops.

I also soon solved gravity because a positive wave that I thought left every proton must react with the negative electrons in atoms, and my school physics gave me the clue of 'back emf'. When you turn the power off in a circuit, the field collapses, forcing the electrons to go the opposite way. It was momentum driving the electron. So then I realised that it is the momentum of these waves that react with the electrons, and because the polarity is different, it sends electrons the opposite way. So I had solved gravity.

Then it was clear that if the wave is space and mass, the question of Dark Matter seemed to be the waves radiating from protons.

It also helped me to clarify 'Duality' because if the wave is mass, then a particle has no mass. It cannot wobble by itself, but it must remain in its wave of space and momentum, so the particle must follow the wave when the wave bends due to interference. And if the atoms around the slit can pull electrons, then the electron wave is changed and can interfere with other electron waves, and the electron will follow the wave.

The only way to be confident that a theory is correct is to check that it can be applied to all aspects of the universe and the only way to do that is to right it all down so that it is easy to cross-check it all, and that is really how this book came about.

The book went through many revisions before I felt it was correct, and I submitted a few unsuccessful papers, then left the project for a few years to work on cancer – because it caused the death of my wife.

Clearly, as I had already decided that the wave of a particle is the momentum and space it needs to move, then the radiation from protons becomes the space dimension of the universe, and because these waves are so dense around a planet or star, they must force themselves apart as soon as they are free of the atoms from where they originated. Thus space is expanding due to the repulsion of proton waves and everything is being pushed away from everything else.

My only remaining question was whether the Higgs Boson that supposedly gave mass to everything and led to so much joy when it was found at CERN, was correct. I now believe that it is nonsense, but I still cannot be sure that the boson that forms my 'proton radiation' is the graviton. But the basic understanding of mass is wrong.

It was equally clear that nobody will believe anything I write unless there is proof, some time ago I contacted Peter Bettels in Germany via his web site to see if we could help each other. In 1997 he had stumbled accidentally on how to achieve levitation, but could not explain it or repeat his success. If I knew how gravity worked, and he had the time and the facility to try it, maybe we would be successful. We failed to achieve it before he had an accident.

So that's where I am now. My book is complete, a paper has been submitted and the project has ended. The problem is that Einstein's Relativity is so well accepted that the entire understanding of the universe is based on it. The academic profession seem to think that those who think Einstein is wrong are nutters! That is the barrier that I am trying to break down. How can progress be achieved while that attitude exists?

I wrote twice to my old university suggesting that they immediately take out a world-wide patent on how to reduce the force of gravity, and that would be my alumni donation worth billions, but I received no response. Such a conceited attitude, or how seriously they believe in Einstein!

So will my theories be accepted? If I am right, someone needs to explain how an engineer, with only school level physics, got things right, and why the profession got so much wrong.

A Short and Simple Message to World Leaders.

We are very fortunate to live on such a beautiful planet. I cannot imagine that there can be a more beautiful place anywhere in the universe. The staggering loveliness of the sea around the Exumas, the massive beauty of Yosemite and the Rocky Mountains, the colours of autumn in Vermont and New Hampshire and even the peace of a warm spring day in a country lane or meadow in England. Please ensure that these lovely things are not destroyed through the greed for money and material things!

Lead the world, not because of the material wealth of your people, or by the strength of your army, but by the protection you give to the planet and the comfort you can bring to those of its people who are less fortunate.

And to borrow similar words to those used by President Kennedy, don't drill for oil or dig for coal because it is easy, find a way to get energy from gravity or some other source, even though it is hard – but may save our planet.

In addition to carbon emission targets there is a need for an International 'Planet Management' organisation that has an over-arching influence on key issues such as rainforests, population size, food availability and poverty.

I think the young people on our planet know these things, and are increasingly voting for leaders who demonstrate such objectives.

Greta Thunberg is certainly stirring things up, but her greatest contribution should be to convince people of her age to limit their number of children to two. Population growth is just as big a risk to lives as pollution and global warming.

My view is that we should cherish and support nature because we depend on it, so we should never destroy any of it.

Chapter Sixteen

WHY DID I DO ALL THIS?

So that is the story of the most difficult problems that I have ever tried to solve. It took only four days to find some of the correct answers and then fourteen years, part time to solve everything, find out where science had gone wrong and to convince the profession that Einstein's ideas may not be the correct solution. So why did I seek to solve the universe?

When I first wrote this book I was living on a lovely quiet island called Exuma in The Bahamas in winter, and on a desert island you either read books or you write them, like Fleming and Hemingway. But a novelist doesn't have to be right; a scientist does! Having the peace on an island, and away from the routine, helped enormously, so it was really just a matter of filling time usefully in retirement and keeping my brain active. Whether anyone ever buys my book is less important, and it certainly wasn't for money!

Footballers can earn double an average annual salary on a Saturday afternoon just by kicking a ball around. The best golfers can earn ten times an average salary in just four days. Pop groups and actors get royalties every time their work is played, so what do brilliant scientists get? The prize for scientists, rather than money, seems to be that you get to keep your job, or possibly get promotion! Where is the Sony World Ranking for Scientists? Where is the glamour of success in this profession?

When you submit a scientific paper for publication there is little reward. As an outsider to this strange environment, that seems to me to be very unfair. Newspaper Journalists, who do little more than make their personal views known to the public, submit articles to newspapers and can maintain a substantial income, and any young girl who spends a night with a known footballer can earn a small fortune just for telling her no-brain story to a newspaper.

And how many names of living scientists do you know? Hawking and… Actually I read that Stephen Hawking was voted second in the UK Good Housekeeping poll of most admired people. That's progress! But it wasn't because of the science community it was Prof. Hawking's initiative to go public with some books.

Newspapers do print the findings of scientific research where it directly affects the public such as in food and health, and corporations or science magazines usually fund this research, but where are the prizes for the big discoveries. Perhaps it is the worldwide Nobel Prize? But we need a UK equivalent, and I am pleased that at last we have the Queen Elizabeth Award for Engineering.

But it still does not seem to me that enough is being done to promote the need for science skills in young people that are essential for the nation's future prosperity. Technology is equally bad, for example; do you know the names of the people who invented the hovercraft, or designed Concorde? Are they millionaires?

Perhaps the scientific profession should have a large 'Promotions budget'? The world is now highly competitive and the Arts are good at promoting their work with several different annual award ceremonies, such as the Oscars, the Golden Globe, BAFTA etc. So why doesn't the scientific profession hold their own annual awards at The Dorchester for the best discovery in each of the main disciplines, including medicine? You may say "The public are just not that interested". I say "We never will get young people interested if we do not recognise and reward achievements".

And the scientific institutions could focus a little more on incentives because money usually provides motivation. If the top fifty physics and science problems were published annually and a £100,000 prize offered to anyone who solves any of them, young people may be attracted to the subject even though their chance of actually winning would be remote.

Aside from poor recognition for the top people, in my view one other reason science is not seen as an attractive career is because the aristocracy, who historically have established the culture in England, consider serving the country in the armed services as the only worthwhile career. 'Safe' subjects such as art or history are their second choice for those who perhaps lack the leadership skills that are essential for the services. Anything technical seems to be regarded as bit 'below stairs'. In fact they could perhaps serve the country better as role models if Prince Harry had decided to study science or engineering, a new trend may have been set. (No way!)

It is just a matter of perception but in the UK, 'Engineers' are seen as people who repair washing machines, and films always show 'Physicists' as being a bit eccentric. The problem lies in the Institutions that control these professions because their standards have not moved with the times. And there is a need for better bridges that link the relevance of these professions more closely to the general public. Scientists are too remote from the general population in their academic institutions and seem to consider the work they do to be far beyond the brains of the man in the street to comprehend when, in fact, it has caused them to lose their common sense and sense of reality. I think this book shows their mistake!. However regardless of cultural matters, young people have adopted computer software development with gusto and considerable prosperity, and this may lead to wider things.

I think I attacked the questions of the universe for the same reason that Sir Edmund Hillary conquered Everest. It became an obsession to do what nobody has done before even though many have said that it was impossible. And once I got a fleeting glimpse of what the answer might be, I could not stop. It became a case of solve it, or go insane trying! Then it became obvious that a lot of work would be required to get all the necessary papers published, for which I would get nothing except a possible mention in newspapers, and no promotion because I do not have a job. My first paper took nearly a year to be reviewed and failed, so I decided to follow Professor Hawking's route, and go straight to the public with a book! Then I started thinking about related subjects and the book got bigger and bigger.

Fortunately, I have long believed that there is no such thing as a problem that cannot be solved. It is just a matter of time and effort.

Einstein had it easy! All he had to do was write two or three theories. I had to prove to the entire profession why Einstein – the believed genius – was wrong, and that many established conclusions were also wrong. But even then, I felt I had to prove my theory by experiment as well. Unfortunately, I lost contact with Peter Bettels. I fear he has had an accident so the experiments are unlikely to be successful.

Summary of fundamental scientific errors

Einstein's Special Relativity is wrong – (time dilation).
Einstein's General Relativity is wrong – (gravity).
The understanding of mass is wrong – (It is the field)
The belief that photons are massless is wrong.
Experiments that claim to prove Einstein's GR theory– (they don't)
The Higgs Field theory is not exactly correct.
Time is misunderstood. It is not a dimension.
The belief that mathematical solutions must always be correct.

My chronology of events.

January 2003 — Read Prof Hawking's book 'The Universe in a Nutshell' and could not agree with any of it, particularly the logic after page 42 when it discussed Ground state fluctuations and Vacuum energy, but it caused me to think about space, time and gravity for the first time.

Dec. 2003 — I had found how space and gravity works, what mass actually was, and the cause of the maximum velocity.

2004 — Wrote my first book 'Time is Running Out' and submitted a paper to 'General Relativity and Gravitation'. Review was favourable but "lacked detail and proof", so rejected.

2006 — Believed I had discovered the scientist's error of particle momentum and spontaneous emission of photons. Submitted improved paper to The Inst. of Physics. Review was that "There is no error". Rejected.

2008 — Discovered proof of one error in terms of Dark Matter and gyroscopic precession from gravity by showing that gravity only attracts electrons in atoms.

2009 — Submitted paper to Physical Review in USA.

	Review favourable but demanded considerably more analysis of past published papers and experiments to prove the errors - a challenge I was unable to meet. (I am not a physicist). Book updated and re-issued.
2011	Added an explanation in the book on why Einstein's General Relativity was wrong, the missing mass, Duality and Young's Slits. Removed my own errors. Then found that the Higgs boson is not radiated from the proton and therefore my 2003 theory on light and Spontaneous Emission was wrong, so scrapped my first book. Submitted an essay to the Gravity Research Foundation. Rejected! Began writing this book.
2012	Realised that the answer was simply proton radiation and not the Higgs boson. Submitted another essay to the GRF explaining everything as described in this new book. I also approached Peter Bettels in Germany and suggested we work together to prove how to achieve levitation. Not good! We failed to cancel gravity and my essay also failed.
2014	Submitted a paper to General Relativity and Gravitation explaining that Einstein had made an error so that Special Relativity was meaningless. It was rejected because it contained an error. Finally realised why experiments to cancel gravity did not work. Released this book in 2015.
2015	Submitted paper to The Gravity Research Foundation explaining gravity, and that mathematics can give invalid solutions. That is why Einstein's Theories are unhelpful and are impossible in practice.
2017	Realised that the equations in Einstein's Special Relativity

	paper are nonsense, and submitted such a paper to General Relativity and Gravitation. Also realised the mechanism of gravity and intend to submit a new paper.
2019	More detailed explanation of Einstein's error in SR submitted as an essay to the Gravity Research Foundation. (Which failed to win a prize!) So, in 2020 I have submitted a more thorough version to the Institute of Physics.

Thank you for buying my book, and please tell all your friends that it has brought a level of happiness to your life that has exceeded all your expectations. You have lost weight. You have never felt fitter. Your sex life has found new energy and your IQ has been raised to 140!

....

My thanks go to the Bahamian people for allowing me to live on the lovely island of Exuma, the makers of big cigars, Imperial College, the publisher lulu.com, Wikipedia, Stephen Hawking, my father who put me through a sound education at Caterham (private) School, and my wife, whose death led me to ask myself 'what's it all about'?

If you enjoyed this book, my second book entitled **'Man Designed Himself'** is now available. It discusses evolution and cancer in the same way as this book discussed the universe, and it contains more surprises! I believe I am on the right track in saying many cancers, particularly breast and prostate, are caused by excess energy in the nerves that reaches us through the eyes. This occurs because we take holidays in sunny countries, we have electric lights at home and we stay up late watching TV, so that the accumulated energy never has a chance to disperse. So wear sunglasses! Skin cancer is the same:- the cells are overwhelmed by excessive sunlight energy. We are electrical machines, and it is the electricity from the sun that allows growth and evolution, but if we get too much, the genes change, and that is what causes cancer.

www.ingramcontent.com/pod-product-compliance
Lightning Source LLC
Chambersburg PA
CBHW070230180526
45158CB00001BA/319